Social Drivers In Food Technology

Vivian-Lara Silva

Social Drivers In Food Technology

 Springer

Vivian-Lara Silva
Faculty of Animal Science and Food Engineering
University of São Paulo (USP)
Pirassununga, SP, Brazil

ISBN 978-3-030-50376-5   ISBN 978-3-030-50374-1   (eBook)
https://doi.org/10.1007/978-3-030-50374-1

This Springer imprint is published by the registered company Springer Nature Switzerland AG
The registered company address is: Gewerbestrasse 11, 6330 Cham, Switzerland

For my lovely family
and all my dear students
Thank you all
for making me look outside the box

Preface

I have been very honored by Professor Vivian Lara Silva's invitation to preface her book "**Social Drivers In Food Technology**," edited by Springer. I have followed Prof Silva's career since she joined FZEA-USP as a Lecturer, when I was Head of her Department, and Vice- and Director (Dean) of the Faculty, but above all, as her colleague, a colleague who saw, from the beginning, the importance of her arrival at FZEA. Since then, I have been sure that she would contribute to making the Food Engineering course at FZEA different from other similar courses in Brazil.

She has been responsible for the undergraduate courses on "Food Distribution" and "Agro-Industrial Management" for the Food Engineering students. In the first course, she teaches, among others, concepts of supply chain and logistics, a novelty in Food Engineering courses in Brazil. In the second course, she addresses the concepts of industrial management, but with a more important human vision, as discussed in this book. I believed that, with her vision of an Agro-Industrial Production Engineer, she was perfectly able to add something more than merely the training of our Food Engineers.

In addition to the aforementioned disciplines, she created discussion groups with both undergraduate and postgraduate students. Besides, she always knew how to motivate them to work as a team. I was always impressed with her ability to "get" students to work on projects. A great and charismatic leader, no doubt.

In addition, she knew how to overcome difficulties, or even turn them into opportunities. So, with a lot of determination and boldness, she decided to do her second thesis, called at USP as "Free Teaching Thesis," necessary to become Associate Professor. In this case, as a candidate, she was her own advisor, that is, there was no such advisor as usual in the regular Doctorate courses. Also, she knew how to orient herself and produced an excellent and innovative work, very different in the context of the Brazilian Food Engineering. In other words, she well knew how to work 'outside the Box', whenever she saw an opportunity.

Because of her second thesis, she published some papers, making herself both known and recognized. That is why it surged the invitation for this book editing, making it simply wonderful, a novelty.

Food Engineering teachers should read and use this book in their classes. To associate the knowledge and reflections presented in this book, makes it necessary with the subjects taught in the hard skills disciplines. In this way, students will be trained with some nuance of soft skills. Nowadays, if food safety and/or security concerns, health issues related to processed foods, environmental issues related to food production and processing (including the *Nexus* food, water, and energy), and entrepreneurship concepts and skills must be considered in the formation of our young students on Food Engineering; ethical and sociocultural issues, or, in other words, knowledge on "**Social Drivers In Food Technology**" must also be taken into account. This book certainly meets with the abovementioned issues. It will contribute to prepare students to face the challenges of the future, which are unknown.

Congratulations to Professor Vivian Lara Silva, for offering us such an interesting and innovative book. I am sure that this book will be a success, a best-seller.

Paulo José do Amaral Sobral
Full Professor
University of São Paulo
Department of Food Engineering
São Paulo, Brazil

At a Glance

The seminal provocation of this book starts from that statement: **social drivers and an unprecedented information problem that is challenging the food industry**. In my personal view, taking us to the best of times. Starting from this scenario, this book aims to inspire you to dive into what is attributed as an opportunity of reinvention of the sector, at the same time of rethinking of its contemporary professional.

The book is a result of five years of dedication, in which several people were decisive, helping, encouraging, and giving me all the necessary support for the publication of this text by Springer.

Expressing my most sincere gratitude and acknowledgment to all of them (as well as to the groups they are part of) will also be relevant for me in order to contextualize the backstage of this book, regarding its nature and inspiration since the beginning of my professional career...

In 2005, with my newly obtained PhD degree, I received from a dear Professor in the Department of Production Engineering at UFSCar, Prof. Miguel A. Bueno da Costa (affectionately called Mian), the suggestion to apply for a position in the Food Engineering Department of the Faculty of Animal Science and Food Engineering at the University of São Paulo (FZEA/USP), located in Pirassununga, São Paulo State. A position that, according to him, had everything to do with what I was working on.

At that time, FZEA/USP was just starting a new Food Engineering program, with a different proposal to the training usually offered in Brazil, inspired by the French school, presenting a vision toward food systems (*filières agroalimentaires*). In charge of this project was Prof. Paulo José do Amaral Sobral, who, for my joy, became one of my mentors and a dear friend of my family. Helping him in the updating of the program under the food system approach, Prof. Decio Zylbersztajn, whose I had the honor of being a student.

I joined FZEA/USP in September 2005, collaborating in the food chain coordination study field and being responsible for the undergraduate courses of 'Agro-Industrial Management' and 'Food Distribution'. Since then I have been very well received, feeling at home and weaving bonds of friendship that I would carry with me throughout my life.

But I have to confess that professionally I felt a little strange and out of place. Looking to inspire me, Professor Paulo Furquim de Azevedo (my advisor in the Master and Doctorate programs) used to tell me about the power of this combination, being in a school of a technological nature with my managerial background, but I did not know exactly how to proceed and explore this apparent opportunity in a research pipeline that would make sense not only to me but that could contribute to the institution, the department and, above all, to the young people in training, my dear students.

A concern that was a landmark in the story of my life and, thanks to other Professors, placed me on a path that I could have never imagined and that culminates in this book published by Springer.

It is important to remember that since 2002, after my exchange period (as Visiting Scholar) at Université de Paris 1 Panthéon-Sorbonne in Paris / France, under the supervision of Professor Claude Ménard, I spent a few years without going abroad. Also, I had started a new professional challenge: to combine teaching, research, and extension activities with the responsibilities of being a mother, with the birth of Catarina in 2007.

Two years passed before I returned to the international scene. It was in 2009 when I had the incredible opportunity of going to Portugal, integrating the mission of FZEA/USP coordinated by Professor Paulo Sobral. A few years ago, we had started the process of internationalization of our food engineering curriculum, counting on the support of CAPES (Brafagri Program) and the encouragement of another Professor that, for my happiness, was also a very dear friend, Domique Colin (Oniris/France).

We were all gathered at the 6th ISEKI-Food Conference 2009, when, among the social activities of the Conference, I had the honor of attending a special dinner in celebration of the retirement of the Professor Alberto M. Sereno (University of Porto). As he was a person loved by everyone, that dinner managed to bring together many important names on the food technology field; that even I, not being from the area (I did not feel yet), was thrilled to be beside those "giants."

And especially because, among the Professors attending the dinner, two of them had given plenary sessions that deeply impacted me during the Conference. They were the renowned Professor R. Paul Singh (University of California, Davis) and Professor Harris Lazarides (Aristotle University of Thessaloniki). Two "giants" in the technology area that in their speeches addressed a systemic perspective in which, according to them, processing should be obligatorily embedded, considering the social and environmental nuances that could no longer be marginalized from the discussion of the technological field.

Their lecture at the time, 2009, was completely out of the curve — out of the box — and more than that, in food engineering in Brazil, there was no such discussion. And then, everything made sense to me, I was impacted.

I came back full of new ideas for new projects. Starting with a conviction: to establish a partnership with those Professors. For my joy, these plans have been achieved. And to my personal satisfaction, Professor Harris Lazarides became a great project partner and a dear friend of my family. Presenting us with two trips to

FZEA/USP, providing us with very rich debates about the socioeconomic environmental context linked to food technology. On his second visit, we received him with our family already extended. Along with me, my husband João Adriano, and Catarina was João Francisco (born in 2012).

Time passed, and a desire to improve my academic career grew within me. Which required me to subject myself to a new exam to become an Associate Professor; and following the rules of the University of São Paulo, the elaboration of a new thesis.

It was here that the guidance of another dear Professor from FZEA/USP, Holmer Savastano Júnior, was decisive in my life. "Vivian, allow yourself to dare, leave your comfort zone trying to develop a thesis with full adherence to food engineering. Learn to translate what you do for the food engineering field." That was the impulse that gave me the courage to take a step further than the scientific trajectory I was dedicated to. Until then, I had mainly been linked to a research agenda focused on contract theory and franchising. And I confess that Professor Holmer deeply impacted me too. Changing me much more than I could imagine.

Closing this historical rescue, almost simultaneously to that advice I have received, Professor Alberto M. Sereno came to spend a year with us at FZEA/USP as a Visiting Professor answering an invitation from Professor Paulo Sobral. And so, Professor Sereno, whose eyes could enlighten everyone around him, particularly warmed my heart to illuminate the path I should follow.

Here I need to mention another key person in my professional and personal life, my mentor and dear friend, and partner in so many projects for so many years, Professor Maria Sylvia Macchione Saes (FEA/USP), coordinator of the research group I participate in, CORS — Center for Organization Studies. With Professor Sylvia, I had the joy of inviting Professor Sereno to give us a lecture on "Food engineering: historical references and future perspectives," discussing main technological milestones in the area of technology and food processing and the opportunity for the sector. During his lecture, all made sense to me as if he gave me the last missing piece of the puzzle I was trying to solve. I had found a new route to follow and dive myself in, and I had found the direction for my new thesis to be presented in the Associate Professor exam.

From here, I started to work, exploring the history of the contemporary industry in order to discuss opportunities for the sector and the impacts on the training of human resources. A few months later, during the 2015 Carnival, I took the courage to send some pages of a first draft to Professor Paulo Sobral asking, "Professor, what do you think about that?". And he answered me, with enthusiasm and encouragement, that it was all I needed to move on and finish this new challenge in my life.

In this process, giving me support and helping me to revise the manuscript, once again Professor Sylvia Saes and the three other dear friends and partners, Professor Carmen Sílvia Fávaro Trindade (FZEA/USP), Professor Roberta de Castro Souza (POLI/USP) and Professor Rubens Nunes (FZEA/USP). The patience, suggestions, support, and friendship from them were essential not only in this process that I report here but for my personal and professional life.

I took the exam in 2016, becoming an Associate Professor at FZEA/USP from there. And all these events in my previous trajectory were turning points in my professional career. Obviously, I brought all my training background, but now having managed to find a line of research to call mine, with adherence to food that positively impacted myself, the institution, and the students. My professional rediscovery, supported by so many special people, gave me a really fruitful research agenda that, since then, I have had the honor of developing alongside great experts in the field.

Following, a series of publications derived from the thesis as well as all my professional rediscovery that I tried to describe here. One paper in particular needs to be addressed, 'Food Industry and Processing Technology: On Time to Harmonize Technology and Social Drivers', co-signed by Professor Paulo Sobral and Professor Alberto Sereno (Food Engineering Reviews 2018), from which, the Springer team got to know my line of research and made me this special invitation for publication.

Other equally important studies were also developed with important names in the area of economics and food technology that life also granted me the honor of having as friends, such as Professor Fausto Makishi (UFMG). Also, to my personal joy, studies that had the decisive collaboration and passion of undergraduate and graduate students, such as Isabella Camargo Fiori, Márcia Kasemodel, Marcus Magossi, Mariana Campos Granado Silva, and Mari Leite Mari Granado. Students were essencial partners in my professional rediscovery, through them, I thank the other students that I have had the opportunity to Interact with.

A network with which I have been also dedicating myself to the discussion of challenges and opportunities that await the contemporary professionals considering the 'outside the box' and the power of soft skills in engineering training. Here, two publications also need to be highlighted: 'Are we doing our homework? An analysis of food engineering education in Brazil' (International Journal of Food Studies 2018); and the series, released every two years, 'Where have you been? Mapeamento dos egressos do curso de engenharia de alimentos da Universidade de São Paulo' (Portal de Livros Abertos da USP, 2019), dedicated to map the trajectory of our alumni. Thanks to the passion and hard work of an incredible group of trainee students of the Project, each edition, we found unprecedented rates among similar studies. In the last edition for instance, available in english, 80% of all food engineering graduates from FZEA/USP were located! From this path, we could better highlight the role of soft skills in engineering education, wich could be empowered through activities out-side the classroom in real-world problems allowing to acquire knowledge to be added to hard skills, composing essential skills for contemporary world professionals.

Other dear students, now prominent professionals in the field, inspired me toward my diving and immersion in the theme of social innovation and social impact businesses — which in the book presented here is materialized in the discussion regarding food ethics. On their behalf, I thank all the other students who directly or indirectly accompanied me on this journey: Lucas Silva Nicoletti, Victor Kim Fukue, and Vitor Vannucchi Ungari.

An agenda that remains alive and that ends up feeding back my trajectory, now with new friends and partners such as Professor Maria Teresa Freire (FZEA/USP), as well as Professor Muriel Fadairo (*Université Savoie Mont Blanc*), Professor Antonio Mulet Pons, Professor Juan Carcel and Professor Neus Sanjuán of Grupo ASPA, from *Universitat Politècnica de València* (UPV).

Speaking of a partnership that allowed me to materialize this book, I also need to highlight my special thanks to Lucas Renato Trevisan (Biosystems Engineering graduate at FZEA/USP, PhD student at ESALq) and Nathalia Fontanari Ortega (Food Engineering graduate student at FZEA/USP). Lucas was instrumental in the entire process of English translation and proofreading. While Nathalia contributed to the graphic editing of the manuscript.

The book presented here is the result of this journey, of this trajectory that, with emotion, I tried to rescue; and that from the names mentioned here I take the liberty of thanking all those who (although not specified) were equally essential in this adventure.

And here, no less important acknowledgments to the Editor responsible for the work, Daniel Falatko, who believed and made the advance and materialization of the invitation by Springer happen along with Deepak Ravi and Cathrine Selvaraj.

My sincere thanks also to the institutions that believed in my work and welcomed me: University of São Paulo and the Faculty of Animal Science and Food Engineering, as well as Center for Organization Studies (CORS) and Group of Studies and Research on Strategy and Vertical Coordination (GEPEC) for providing a rich environment for the development of this study. Thanks also to the Brazilian National Council for Scientific and Technological Development (CNPq) for financing the project that led to the preparation of this book (Universal Project, Process 407 498/2016-8).

Teams and support that are still essential at the beginning of a new cycle that I started at the same time that I materialized the dream of editing the book by Springer: a new and ambitious program, All 4 Food. An Academy Active and Collaborative Connection Program with multiple stakeholders in the food and beverage innovation ecosystem. It is an honor coordinating this program, with more than 22 research, extension and teaching centers that are connected to some of the leaders in the processing sector, NGOs, Accelerators, Associations and Student Institutions. Still, a multidisciplinary network, composed of experts and researchers from the great areas of business and food technology. We came with the proposal of adding value by monitoring and discussing the dynamics of innovation in food in the light of sustainable development, to promote new business and creative solutions in collaborative innovation projects, while illuminating new challenges to science and stimulating a new mindset in the training of human resources.

And finally, but naturally not least. Actually, quite the contrary.

The whole process reported here was encouraged and supported by my family's patience, for the days, weeks, months that I was reclusive, working, missing their interaction. Always with the endless support, patience, and love of my greatest partners of all – my husband, João Adriano Rossignolo, our children, Catarina and

João Francisco, together with my fathers, Maria and Eurípides, brothers, brothers in law and nephews.

And, obviously, encouragement and support from my students, that with a sparkle in their eyes challenged and fed me with cases, perspectives, and dreams, believing and betting with me on the power of social drives in food technology.

Book Proposal

The motivation for this book is based on the author's experience of 15 years in food engineering at the University of São Paulo (USP), one of the most prestigious institutions in Brazil with a great international reputation.

One of the author's main objectives as a Professor is to stimulate students to perform a critical analysis of food processing technology, considering the incorporation of elements that intersect the technological development arising from the contemporary context in which the food sector is inserted.

The main message the author seeks to explore in the classroom and in her work as a researcher is: it is time to harmonize product and process engineering with relationship engineering from farm to fork.

Playing with a classic food science and technology term, this message refers to the current challenges or, as she prefers to understand and instigate her students, the opportunity for both the sector and its professionals, considering the existence of new food consumption drivers, i.e., social drivers.

Going beyond consumer's physical health (usually associated with technological advances in product and process), also considering the health of the society in which the food company is inserted (in terms of construction, and maintenance over time, of a real and healthy history among the processor company and its partners). Which, in practical terms, reveals new opportunities that can no longer be overlooked by the food industry, and require a repositioning of the processing sector, in terms of understanding that more than innovating in product and process, the opportunity is also in preserving and enhancing attributes derived from the relationships established throughout the value chain in which the food industry operates.

Look beyond technology. Look outside the box. And from this exercise discuss the opportunities for the food processing sector and its professional, which involves (re)thinking bases of product and process innovation, knowledge generation, and human resource training. This invitation has more recently been highlighted and evidenced by some of the most important authors in the food science and technology field, but still remains unexplored in terms of the discussion of its economic and managerial implications.

This context reinforces the author's motivation to explore this perspective with her students and, at the same time, provides the opportunity to transcend her immediate limits and take this discussion beyond teaching—providing, among other opportunities, Springer's invitation to publish the project under review, i.e., a textbook about the *Social Drivers in Food Technology*.

In essence, the textbook will consist of a scientific study based on an extensive literature review, including scientific publications and organizational reports from food processing companies—linked with a fresh and attractive approach by the discussion of case studies, customized for the specific purpose of the publication, enriched with images and illustrations collected by the author. Six food sectors will be highlighted in the textbook: chocolate, coffee, yogurt, juice, baby food, and snacks. These sectors were deliberately chosen in order to illustrate the dynamics through which the food processing technology has passed.

Despite not being an economics textbook, it addresses the economy applied to the technological universe of food processing. This multidisciplinary approach is motivated by the opportunity to contextualize the technological debate regarding the strategy and organizational economics fields, contributing to the generation of knowledge and the training of human resources. Proposal directly influenced by the author's research interests. With a background in industrial engineering, she has expertise (Ph.D. and Postdoc) in the value chain and organizational economics, with a particular interest in issues related to food ethics and corporate social responsibility in the food industry.

Created and designed for academic purposes, the textbook proposal is comprised of two parts (Industrial Approach – see Chapters 1 to 3; and Economic Framework – see Chapters 4 to 6), which can be read and explored in sequence or not, according to the preferences of the reader or even according to the dynamics idealized by the professor in classroom activities. These two parts consist of six chapters, as detailed next.

A pleasant and thought-provoking reading is expected—informative and provocative for students in training. Theoretical discussions are illustrated and enriched with practical and real cases of study, and yet with the provocation of showing that social drivers already guide—or at least influence—the orientation of the industry worldwide, including, and particularly, in developing countries—a fact that justifies the inclination towards Brazil, in addition to being the author's homeland and place of residence. More than convenience, we also sought to innovate by bringing Brazil as a highlight to illuminate the discussion beyond the regions usually considered in the studies. And more than that, we invite new studies.

Contents

About the Author

Vivian-Lara Silva Born in August 1975, I confess that only recently I have understood my real connection with my own name. Vivian, from the Latin *Vivianus*, derived from the term *vivus*, which means "alive." It denotes vitality and joy…

Personal characteristics that are present in what I am and what I do, in my essence as an equilibrist in the challenging and delicious tasks of being mother, professor, and researcher, all together and mixed up, not necessarily in this order…

For my joy, inspiration, and responsibility, I am Catarina and João Francisco's mother; and I come from a family of educators. Married to a professor, I am also the daughter of educators, sister of educators, and sister-in-law of educators. With several uncles and cousins that are also educators. This helps to explain my passion and enthusiasm for education in the process of social transformation.

I also love to eat … and I like (I cannot resist) any kind of innovation in food. According to Catarina and João Francisco, my favorite dish is the "full plate"; and one of my favorite hobbies, guess what? It would be eating while we talk about the stories behind the food. And they are totally right.

Discussing and learning how food has arrived at our table is something that fascinates and instigates me. Addressing challenges and opportunities for all stakeholders participating in the value chain involved in the transformation of inputs into food products (semi)ready for consumption.

A subject that I believe to be of total relevance for the education of my children as citizens. In the same way that it is fundamental for my students, as challenged professionals — or, as I prefer to say, empowered — by the contemporary context in which food processing and marketing are found. At the same time, it seems to be an inexhaustible source of inspiration to continually feedback into my research agenda. All together and mixed up…

Evidently, all of my emphasis on food also comes from my background and professional field of work. *Food Industrial Engineer by training; Business Economist by specialization; and Food Engineer by heart.*

That is how I understand myself professionally; and at each new academic course, I introduce myself to my dear students. I am graduated in Agro-Industrial Engineering, with a Master's and a PhD in Industrial Engineering (both of them on relationships governance and coordination of food value chains) and a Postdoctorate in Food Business Economics. And since 2005, I am full-time faculty at the Food Engineering Department of the Faculty of Animal Sciences and Food Engineering at the University of São Paulo (FZEA/USP).

In this personal and professional trajectory, and thanks to the daily contact — and learning — with my students, colleagues, and family, I found myself attracted by the magic of the discussion about food processing from the perspective of the new consumption drivers; and the obligation to (re)think the differentiation bases of food technology, considering innovation in the organization of food systems. In my personal opinion, a path with no return. There is no way back in fact. Here we are in "the new norm"!

Well, some might call me a dreamer and romantic — another characteristic that makes sense to associate with my personality. Anyway, in the prime of my 45 years, I prefer to continue following this story and trying to assume my role of responsibility in the transformation that is happening.

Social drivers in food technology. Food ethics. Valuing the history behind the transformation and commercialization of the food that is on our table. Reinventing itself as a sector. Understanding that technology is no longer self-sufficient. Whether due a genuine social impact action, or in response to competition.

And from this cycle, redefining both the research agenda and teaching activities, rethinking the training of contemporary professionals and the education of our children in order to empower them in terms of social drivers – encouraging them to identify and explore new opportunities for innovation in food, assuming their role in the transformation that is witnessing.

All together and mixed up! Mother, professor, and researcher, not necessarily in this order. But continually feeding back the joy of life and the belief in humanity and collaborative networking.

I hope that all of this can also make sense to you and that we can continue together, building and following everything that lies ahead.

Chapter 1
The Story Behind

> *"When it comes to food engineering you have to look outside the box. There are lots of opportunities right at the boundary and you need to make sure that you are prepared for the cutting-edge work."*
>
> Paul R. Singh (2012)

Abstract The history of the food industry, from its seminal conception in the early 19th century to the current days, represents an instigating story that brings us back to an important statement. More than creating (new) textures, the ongoing opportunity is also in preserving and enhancing attributes derived from the relationships established throughout the value chain in which the food industry operates from farm to fork. Starting from this picture, the main goal of this chapter is to address the past and present of this dynamic, including a discussion on what is expected for the next chapters regarding both sector and technology of food processing.

Keywords Food processing technology · Processed food · Food industry history · Food science and technology knowledge · Paradigm shifts · Food market · Food trends · Shared value · Health and Wellness · Social drivers · Human resources training

1.1 Shaping the Future

Processed food represents a symbol of contemporary society. At the beginning of the first decade of the 2000s, more than half of the food volume sold worldwide was already processed (IMAP 2010). Despite it all, consumers seem to not easily understand the benefits associated with food processing.

© Springer Nature Switzerland AG 2020
V.-L. Silva, *Social Drivers In Food Technology*,
https://doi.org/10.1007/978-3-030-50374-1_1

Have you ever tried to Google 'processed food' for instance? Among the answers, a huge amount of websites spreading reasons to avoid consuming processed food. Take for instance the 'real food movement', and its emblematic phrase *"don't eat anything your great-grandmother wouldn't recognize as food"* (Pollan 2009: 148). This view invites us to the consumption of products that are considered 'unprocessed such as organic and whole food categories[1]'; been 'processed food' is understood as one of the greatest villains of our time, a major cause of a global obesity epidemic and associated with diseases such as diabetes, hypertension, cardiovascular problems, high cholesterol, and so on. As a result, worldwide, consumers begin to change their eating habits, impacting on the composition of their cart at the supermarket, in terms of brands and product categories (Nunes et al. 2019; Silva et al. 2019). See more details in the ***Box Magazine cover***.

The controversy goes even further.

Food production and distribution consume a considerable amount of energy, contributing significantly to global greenhouse emissions, as well as to water consumption and waste generation (Defra 2017; Silva and Sanjuán 2019). At same time food industry is also usually associated with complaints about food tampering scandals, market concentration, unilateral relationships with its suppliers, and harmful environmental practices.

In contrast, it's important to consider that, not so long ago, the consensus was completely opposite to this: processing was understood as a hero, helping to save humanity. Indeed, it was due to the evolution of the sector that humanity has had the opportunity for a regular access (volume and periodicity) to a food that is microbiologically safe (food safety) and, at the same time, passive to storage (shelf life); and so with access to a food product with taste, pleasure, and nutrition, as well as health and well-being claims. A context that deserves a lot of caution in its analysis though. Intending to warm up the discussion, take a look at the ***Boxes My, your, our fault***, and ***Controversy on the plate.***

So, what happened then?

Adding some extra flavor to the discussion, in spite of all technological progress, it is quite clear the sector has not proven to be equally efficient in the articulation with different stakeholders (Silva et al. 2018).

And now, one of its challenges is (re)building consumer trust (The Center for Food Integrity 2018). See the ***Box Trust alert***. Another challenge is to understand that opportunity is found outside the box (Singh 2012), breaking the usual technological bias.

A new paradigm should shape food research and development: food systems and the understanding that the perceptions consumers have about food value and quality is-also-based on how it is made and where it comes from (Lopez 2014).

What about next?

Here we find a consensus: the food industry has the challenge of reprogramming itself once again. See **Box *Glossary***. Reinvent itself considering that consumption patterns have truly changed towards social drivers in food technology. What requires

---

[1] About this urban legend, see **Box *Controversy on the plate***.

harmonizing product and process engineering in alignment with relationship engineering set up by the food industry along its value chain. This tour starts next with a review about the origin of food industry.

Magazine cover

The challenges became evident at the beginning of 2017 when the big players of the food industry announced changes in global consumer habits. In their shopping carts, healthier products. *"There was a pattern in 2017 within the food and beverage sector that the whole industry took too long to recognize, but it is now in their sights"*. Nestlé and Danone have registered slower expansion through the last 20 years. In turn, Campbell Soup observed a 2% drop in sales in the fourth quarter of 2017, due to the low demand for its traditional soups in North America. Kraft Heinz has presented the seventh consecutive drop in sales, embittering a 1.1% fall in the USA market. While Coca-Cola has reported its biggest drop in soft drink sales since 1986 (a 31 years period). This scenario gains prominence in social media and leads the sector to manifest and, more importantly, to reprogram itself. According to Mark Schneider, CEO of Nestlé, the trend of quitting 'super-brands' has come for good (Nicolaou 2018).

My, your, our fault

The world lives one of its greatest food paradoxes. While more than a billion people are starving, health problems attributed to unbalanced nutrition are a major cause of death, regardless of communicable diseases. About 1 billion people worldwide are obese or overweight, causing obesity and cardiovascular diseases. Sugar, salt and fat 'invisible' to our eyes, but excessive in the formulations, are among the main responsible for what experts define as an ongoing world's true pandemic.

This phenomenon is global, affecting all countries, including developing economies. In Brazil, for example, the situation is alarming. The Brazilian population, especially children, does not stop gaining weight. Among adults, 51% were overweight in 2015 (in 2006, the rate was 43%), from which 17.5% were considered obese (11% in 2006). Among children from 5 to 9 years old, one-third were overweight in 2015. If this circle is not interrupted, Brazil will present the same rates seeing in the USA, where 75% of the adult population is overweight and another 30% is within the obesity range. However, it's important to highlight: this phenomenon is related to changes in eating habits, but not restricted to it.

Food industry and fast-food chains are not responsible alone nevertheless. *"The responsibility is (also) ours. Accepting this fact is the first step in facing obesity"* (Varella 2015). Which means that dealing with obesity and cardiovascular epidemics requires calling up a task force of the different stakeholders involved: (1) Individuals, sharing knowledge to raise awareness about the benefits of adopting healthy habits, which includes physical activity and balanced food plans; (2) Society, fomenting discussion on the problems related to obesity, suggesting public policies and educational programs; (3) Government, performing control and inspection; and (4) Industry, reviewing its products and services, also investing in consumer education and obesity treatment programs.

Controversy on the Plate[2]

The global packaged food market will be over US$ 2.2 trillion by 2021 (Euromonitor International 2016), underscoring food industry among the main industrial sectors, but also places it at a forefront position in terms of its social and environmental burdens (Silva and Sanjuán 2019). Nevertheless, some considerations on the critical view of processed food should be pondered, remarkably concerning what is usually marginalized from the discussion attributed to food processing. In this respect, Silva et al. (2018) and Silva and Sanjuán (2019) discuss four main points, to which another three will be added here.

First, "It's difficult for consumers to accurately define the concept of processed food" (Wan 2012: 73–74). On a regular basis, other aspects end up being considered, such as cultivation methods and quality of the relationships practiced by the company towards its suppliers. Organic products, halal, kosher, vegan, Demeter, coming from fair and solidarity trade are often interpreted as comparatively less processed and, consequently, healthier. Far apart from this 'urban legend', these categories refer to questions that transcend the industrial processing itself, in terms of the chain of steps employed, or the characterization (level of transformation) of each of them (Silva et al. 2018). In fact, halal and kosher mean, respectively, proper food for consumption according to Islamic and Jewish principles. In turn, vegan food denotes a vegetable-based diet, without any animal-based food. And lastly, to look only at the examples mentioned above (although the list could be much longer) Demeter food represents an international food seal with the concept of biodynamic agriculture (which can be understood as an evolution of the organic principle).

Second, on the basis of all food trends, there is a food processing company. Any company that uses some level of processing (simple or complex, unitary or sequential operations) should be understood as a food processing industry. The acquisition of animal or vegetable raw materials, certification of

---

[2] Based on Silva et al. (2018).

origin, pre-processing (cleaning and selection), packaging, distribution and marketing, represent activities performed by food processing companies.

*Third, **the food industry engages an indispensable economic and social role***. Statement related to the scope of the food industry, and mainly to its relevance in the generation of divisions and employments, direct and indirect, revealing itself as one of the main sectors for the growth of the world economies. In fact, *"food industry is bigger than people think"* (Saguy et al. 2013).

Fourth, rescuing Silva et al. (2018), ***"all-natural, (not) always safe"*** (Wan 2012: 56). Findings that weakens the relationship between *in natura* and safe products. Usually, the population puts health at risk when indiscriminately consuming herbal infusions. *"Because they consider plants as something totally natural, they think (population) that there are no risks"* (Silveira 2008). However, the presence of toxins in the composition of *in natura* products, including medicinal herbs, requires meticulous control over the form of preparation and consumption (dry or fresh herbs, leaves or flowers), cultivation (soil, use of agrochemicals, harvest season, etc.) and form of processing. Such carefulness is essential to consumer safety (Silveira 2008; Segatto 2010) but the population, in general, is not aware of or underestimates it. A discussion that suggests a new reflection aspect of the critical bias attributed to food processing.

Fifth, ***"Not all processed food is bad for us"*** (Wan 2012: 3). In fact, *"if you teach a person to process food, you can feed a village"* (Nelson 2013[3] see in Weaver et al. 2014). Processed food takes a priority role on food safety according to the current definition by FAO (Food and Agriculture Organization of the United Nations), which specifies *"food safety and quality for all"* as a function of the binomial: quantity (accessibility, availability and regularity) and quality (microbiological and nutritional aspects). A classic example is pasteurized milk. Pasteurization is vital to the removal of bacteria that can culminate in serious foodborne diseases. Another example is food freezing, which, coupled with convenience in preparation and consumption, provides a source of food energy, preserving vitamins and minerals (Wan 2012: 3–4).

Sixth, ***"food processing is necessary"*** (Floros et al. 2010). *"In addition to the economic, political, and social reforms, more widespread use of methods to preserve, store, and distribute foods in developing countries can help to alleviate some problems of under and malnutrition due to poor food distribution" (Smil 2001; Conway 2012). Also, improvements in processing capacity (e.g., to safely preserve, store, and transport food) are needed to reduce food waste and to better ensure an adequate food supply as the world's population grows (FAO 2018). Nutrition and food scientists, public*

---

[3] Philip E. Nelson was awarded the World Food Prize in 2007. Recognized as the Nobel Prize in Food and Agriculture, this distinction aims to recognize and inspire innovative achievements in order to guarantee world food security. More information at http://www.worldfoodprize.org/.

health professionals, agricultural economists, and other professionals dedicated to meeting the food and nutritional needs of people around the globe recognize that fresh local foods cannot meet all their nutritional requirements" (Floros et al. 2010).

This discussion forms a base to the relevance of the role that processed food present also in terms of nutritional security (Weaver et al. 2014). A research based on the values used by the US nutrition chart-National Health and Nutrition Examination Survey 2003/2008 shows, for example, that processed food plays a dual role in daily diets: whether to encourage the necessary minimum consumption of certain important nutrients or to contribute to the reduction/control of those harmful to the organism when ingested in excess (Weaver et al. 2014: 1525). Based on the daily reference values for the American diet, processed food contributes to the intake of 55% fiber, 48% calcium, 43% potassium, 34% vitamin D, 64% iron, 65% folic acid and 46% of vitamin B12. On the other hand, efforts developed by the industry have been contributing to the reduction of harmful nutrients present in the food. In the US, processed foods contribute to 57% of calorie intake, 52% saturated fats, 75% added sugars and 57% sodium (Weaver et al. 2014: 1525). This study corroborates the vision that health aspects (and the obesity pandemic, **Box My, your, our fault**) transcend the processing level of ingested foods, maintaining direct relationships with the food balance (Weaver et al. 2014). Indeed, one of the central points for the effective improvement of the American diet (which can be extrapolated to a worldwide point of view) involves consumer education about all the nuances of this discussion (Weaver et al. 2014).

Making the problem even more complex, *"consumers don't know what information to trust"* (Wan 2012: 42). What leads to the sharing of a subjective perception and distorted information concerning processed food and their role in relation to health (Wan 2012; Saguy et al. 2013). Suggesting the challenge encompass an effective communication between all the different stakeholders involved (Lazarides 2012, Weaver et al. 2014, Silva et al. 2018), looking for to attenuate the language barrier that seems to exist among professionals and consumers; government and industry; and also between firms in the same processing segment (Wan 2012).

And last, but not least, *seventh*, according to Silva and Sanjuán (2019), **food processing and environmental impact do not necessarily have positive relationships**. In fact, energy efficiency, heat recovery and reverse logistics, for instance, contribute to increasing the environmental performance of food processing, reducing its impact on the environment. Of course, this statement could be not generalized, particularly when comparing industrial processing with homemade cooking. However, some results bring suggestive insights in directing us to open our minds regarding processing. For example, Calderón et al. (2018) show that large-scale systems (ready meals industry and catering companies) incorporate measures aimed at energy saving and waste reduction and, thus, can offer a better environmental performance than small-scale systems, such as eating in restaurants or even cooking at home (Silva and Sanjuán 2019).

Trust Alert

The Center for Food Integrity (2018) alerts to a dangerous moment the food industry is living related to a gap between responsibility and trust regarding consumers' views. According to them, the main issue is when you're held responsible but are not trusted to deliver, such as federal regulatory agencies and food companies. The consequences of this gap are: Advocacy for more oversight and regulations (a) Rejection of products or information (b) Looking for alternative and perhaps unreliable sources.

Source: Center for Food Integrity (2018). Retrieved from: http://www.foodintegrity.org/wp-content/uploads/2018/01/CFI_Research_8pg_010918_final_web_REV2-1.pdf

Glossary

The terminology of the food industry is used as a reference to the complex economic sector involved in the industrial transformation of raw material of animal and vegetable origin into a final product, food or drink, ready or semi-ready for consumption. Due to the complexity assumed here, the food industry deals with a mix of processing technologies and different economic sectors, broken down into different levels of industrial transformation. As will be shown, the conception of this macro and complex vision is due to the sector's own development history.

1.2   The Beginning of Everything

The origins of the industry-as we know it today-are intrinsically associated with war interests, i.e., processing of agricultural inputs for ensuring integrity (food safety) and stability (shelf life) of a storable food source.

For that purpose, in 1809, Napoleon Bonaparte awarded Nicolas Appert a prize of 12,000 Francs for developing a technique that consisted of the thermal treatment of a hermetically packaged food at water boiling temperature. See *Box Appertization*. From then on, processed food businesses started to emerge all over Europe.

But it was only in 1864 that Louis Pasteur presented the scientific foundation of Appert's great realization. As discussed by Silva et al. (2018), Pasteur was invited to solve the problem of early acidification of wine due to uncontrolled fermentation. And this was how he observed that the deterioration of wine was caused by the action of microorganisms. Pasteur benefited from the original design made by Appert, proposing thermal treatment as a means of decontamination. This application, originally intended for the French wine industry, was rapidly disseminated to other sectors and countries, receiving Pasteur the historical recognition (Latour 1984, 1988). As a result, what was previously known as 'appertization' entered history as 'pasteurization' (Silva et al. 2018).

This is usually understood as the origin of the contemporary food industry, which benefited from the context of two revolutions, agricultural and industrial, stimulating new technological developments, and so the development of new products and businesses, among them some of the most important contemporary brands of processed food (Silva et al. 2018). The chronology of this story is presented in the Food Industry Timeline (Fig. 1.1).

According to Silva et al. (2019), war interests led to both the blossoming of the industry in the early nineteenth century, and to its first boom at the end of the same century, extending through the first phase of the twentieth century. This culminated in an important milestone in the seminal development of the sector: attending war demands, *i.e.,* processing to ensure food availability and integrity. While other attributes, such as sensory (taste and pleasure), were for long marginalized. *Boxes Convenience off the menu* and *How it all started* rescue central facts about that.

> **Appertization[4]**
> In its original conception, the technique developed by Appert was based on the use of glass pots, capped with cork stoppers and sealed with wax. Appert attributed the care taken in the bottling process as the decisive factor for prolonging the preservation of the food. In 1809, Appert published the book *"L'art de conserver, pendant plusieurs années, toutes les substances animais et vegétales[5]"*; and his technique spread all over Europe. Two important

---

[4] Silva et al. (2018).

[5] In English, "The art of preserving, for several years, all animal and vegetable substances".

advances contributed to this process. The first one was registered in England, referring to the adaptation of Nicolas Appert's technique for metallic packaging, cheaper and more resistant than the glass originally used. Progress idealized and patented by Peter Durand in 1810. The second, was the autoclave, equipment that would allow food sterilization. Event patented in France by Raymond Chevallier Appert, Nicolas Appert's nephew, in 1852.

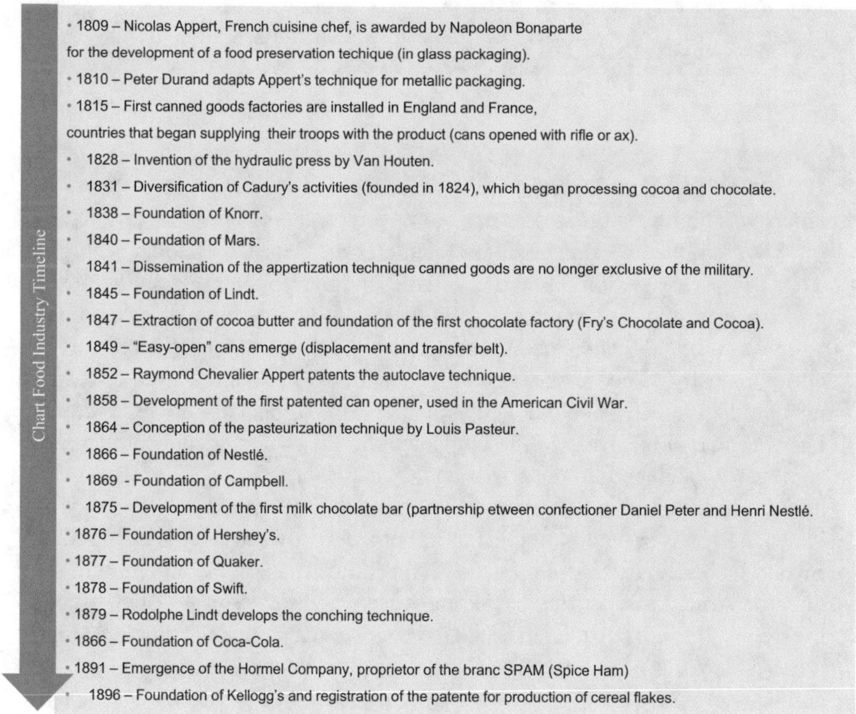

- 1809 – Nicolas Appert, French cuisine chef, is awarded by Napoleon Bonaparte for the development of a food preservation techique (in glass packaging).
- 1810 – Peter Durand adapts Appert's technique for metallic packaging.
- 1815 – First canned goods factories are installed in England and France, countries that began supplying their troops with the product (cans opened with rifle or ax).
- 1828 – Invention of the hydraulic press by Van Houten.
- 1831 – Diversification of Cadury's activities (founded in 1824), which began processing cocoa and chocolate.
- 1838 – Foundation of Knorr.
- 1840 – Foundation of Mars.
- 1841 – Dissemination of the appertization technique canned goods are no longer exclusive of the military.
- 1845 – Foundation of Lindt.
- 1847 – Extraction of cocoa butter and foundation of the first chocolate factory (Fry's Chocolate and Cocoa).
- 1849 – "Easy-open" cans emerge (displacement and transfer belt).
- 1852 – Raymond Chevalier Appert patents the autoclave technique.
- 1858 – Development of the first patented can opener, used in the American Civil War.
- 1864 – Conception of the pasteurization technique by Louis Pasteur.
- 1866 – Foundation of Nestlé.
- 1869 - Foundation of Campbell.
- 1875 – Development of the first milk chocolate bar (partnership etween confectioner Daniel Peter and Henri Nestlé.
- 1876 – Foundation of Hershey's.
- 1877 – Foundation of Quaker.
- 1878 – Foundation of Swift.
- 1879 – Rodolphe Lindt develops the conching technique.
- 1866 – Foundation of Coca-Cola.
- 1891 – Emergence of the Hormel Company, proprietor of the branc SPAM (Spice Ham)
- 1896 – Foundation of Kellogg's and registration of the patente for production of cereal flakes.

Chart Food Industry Timeline

Fig. 1.1 Chart Food Industry Timeline. (Source: Silva et al. 2018)

Convenience off the menu
It is curious to think that even though the convenience attached to canned food is part of the seminal origin of the food industry, these aspects were not considered and valued during its development. As discussed throughout this chapter, the target was to make available an inexpensive source of food energy, ensuring shelf life, even after long periods of storage.

Everything else was ignored, including nuances that today are easily associated with processed food, such as flavor and even the practicality expected
in the packaging used. Going back to the chronology of the development of
the sector, it was just more than three decades after the first canning factories
in Europe (as indicated by the Chart Food Industry Timeline, Fig. 1.1) that the
easy-to-open cans appeared (1849); and another decade more for the patent of
the first can opener (1858).

How was it done before?

The packaging had instructions:

"Open with bayonets or smash them open with rocks!"[6]

How It All Started: Knorr[7]

Knorr was born in southern Germany, in a town called Heilbronn, originating
in 1838 from the experiments of its founder, Carl Heinrich Theodor Knorr, in
a tentative to dry seasonings and vegetables in order to minimize seasonal and
climatic effects on the regular supply of food. With this main goal, the company was originally dedicated to the production of dehydrated chicory for the
coffee industry.[8] The business continued with his sons and started releasing
new products, such as dehydrated soups, in 1873, especially the product that
later became the flagship of the company, the Erbswurst, a concentrated pea
soup paste, launched in 1889, that could be consumed directly or as a soup,
after hydration. It is noteworthy that prior to this product, in 1885, the company started its internationalization process, initially in Austria and
Switzerland. The true development of the company drew the attention of the
German Armed Forces, due to the quest for war strategy in logistics to feed
the troops. Interests that led the state to take over the company from 1899 until
the end of World War II, when the Knorr family regained control. In 1912, the
company provided another innovation in the processing of vegetables, launching the first bouillon cube.

---

[6] See at https://en.wikipedia.org/wiki/Canning

[7] Based on information removed from official company website, see at https://www.knorr.com/us/
en/about-us/brand-history.html

[8] Historical reports suggest the practice of adding dry chicory into the coffee roasting process as a
strategy to lower the cost of the final product.

How it all started: Nestlé[9]
The history of Nestlé begins with the search for a solution to child malnutrition problems by the pharmacist Henri Nestlé. His purpose was to develop a formulation composed of staple foods, such as milk, sugar and wheat flour, which, when dissolved in water, would result in a product with high nutritional value. Henri Nestlé benefited from previous innovations in order to optimize the transportation and preservation of fresh milk, which used to be transported in oak barrels, potentializing its deterioration. The main contribution came in 1856 from dehydration experiments conducted by the American Gail Borden. Prior to obtaining milk powder, the process generated a concentrated product-which is also called evaporated milk, referring to a product with a reduction of about 60% of water. The dissemination of this technique occurred from 1861, with the American Civil War, the moment at which these products proved to be strategic to feed the troops in combat, given the longer conservation time and logistical ease. In 1886, the chemist Franz von Soxhlet, a studious on dairy products, idealized the application of pasteurization in milk. But it was only after 1920 that the production of pasteurized milk gained expression; directly influencing Henri Nestlé and led to the development of Nestlé Milk Flour in 1866. A year later, large-scale industrial production began, a milestone in the creation of Henri Nestlé Society, followed by the production of condensed milk in 1874. In the following year, Henri Nestlé sold his business to a group of Swiss entrepreneurs for one million Swiss francs, a carriage and a few horses. It was created, then, the *Société Anonyme Farine Lactté Henri Nestlé*. In 1898, Nestlé bought a Norwegian condensed milk company, being its first acquisition outside Switzerland. Two years later, the first factory in the USA was inaugurated. With the outbreak of World War I, demand for food increased, especially in the form of government contracts, and by the end of the conflict, Nestlé's production had more than doubled.

How it all started: Nescafé[10]
A curious fact marks the development of Nescafé-another flagship product of the Nestlé (extensively used in the war fields)-and the Brazilian market. In the 1930s, after a super coffee harvest production that dropped the prices of the commodity (from 1931 to 1938, 65 million bags of coffee were destroyed),

---

[9] Based on 'Nestlé. 1866–1905: The pioneer years'. Available in https://www.nestle.com/aboutus/history/1866-1905.

[10] Nestlé.   1866–1905:   The   pioneer   years.   See   at: https://www.nestle.com/aboutus/history/1866-1905

the Brazilian government suggested the company to develop a product that could preserve the coffee similarly to what was made with milk powder. It took 7 years of research until the conception of adding carbohydrates to the raw material to conserve the aroma and the natural taste of the coffee. The result was a soluble product, that could be consumed simply by adding water. The company then assembled a large-scale production line for coffee extraction and grain spray-drying to produce the new product in its factory in Switzerland. The new product was introduced worldwide as the first instant coffee in the market. In April 1940, Nescafé was already present in more than 30 countries. In 1942, with the expansion of the American army around the world, the product became part of the daily diet of American soldiers (more than ¾ of the Nescafé's world production was consumed in Switzerland, the United Kingdom and mainly the United States during World War II). In 1952, a new process innovation: made with 100% roasted beans; and later, in 1966, a technique for preserving the original coffee aroma using 'freeze-drying'.

How it all started: Campbell's[11]
It all started in 1869, in the town of Candem, in the American state of New Jersey, with a business derived from the partnership between a fruit merchant, Joseph A. Campbell, and a can manufacturer. The company did not last long, but Campbell remained ahead of the brand and business, which made history almost three decades later, in the year of 1897, after innovating in their tomato soup manufacturing process, a product that was incorporated into the portfolio of the company in 1895. The innovation consisted of processing concentrated soup, reducing packaging, transportation and storage, reverting to highly competitive prices (about 10 cents a can). The characteristic red and white of the brand only began to be used in 1898, in reference to the colors of the University of Cornell football team (Cornell Big Red). In 1902, 21 different varieties of soup were already being sold, and by 1904 sales reached 16 million cans. In 1918, due to the great demand to supply the American troops during World War I, the company launched the vegetable soup with meat, presenting high nutritional power. In 1922 the company was named Campbell Soup Company.

---

[11] Inspired by official company information obtained from https://www.campbellsoupcompany.com/about-campbell/

How it all started: Kellogg's[12]

Dr. Kellogg and his brother, Will Keith, performed a number of experiments aiming at balanced nutrition for their patients. In 1894, while cooking wheat, he accidentally roasted the dough, resulting in a product that fell into flakes after being pressed. Making the story even more exciting, records suggest that the cooking and pressurizing method created was improved using an army cannon converted into a pressure cooker, which made the cereals literally explode, producing a crunchy texture. In 1896, the Kellogg brothers obtained a patent for the technology developed. Perceiving the opportunity, Will Keith bet on scaling production and improving the product, and, in 1898, he conceived the formula of Corn Flakes. In 1902 the brothers split the company, and the business continued under Will administration. In 1906, he founded the Battle Creek Toasted Corn Flake Company, which in 1922 became the Kellogg's Company. In 1909, the company was one of the forerunners of sale coupons. And in the 1920s, the company began to stamp on its packaging the signature of its owner, as a strategy to differentiate its product from the competition.

How it all started (Cont.): Coca-Cola[13]

In 1886, John Stith Pemberton, a pharmacist from the city of Atlanta, Georgia/USA, created a tonic drink to relieve headaches and stomach pains based on extracts of coca leaves and cola nuts, a stimulant with high caffeine content. The product used to be stored in wooden barrels painted in red, influencing the color that later would help to personalize the brand. Carbonated water was added to the syrup, and the drink was commercialized in his own pharmacy, for five cents (USD) per cup. In 1891, with the business not thriving as idealized, the formula was sold to another pharmacist, Griggs Candle, for about twenty-three hundred dollars.

It was then observed a revolution in how to promote the product, such as discount coupons for consumers and the offer of gifts, with the Coca-Cola brand stamp for distributors. Shortly after, a merchant from the state of Mississippi proposed to market the product in glass bottles.

---

[12] Kellog's: time line. Available in http://www.kellogghistory.com/timeline.html.

[13] Based on Coca-Cola, Company history, available in https://www.coca-colacompany.com/company/history

This proposal was sub-appreciated by Candle, who, for the symbolic value of a dollar, sold the exclusivity rights of bottling and selling the beverage. By 1895, Coca-Cola was already sold throughout the United States, with three bottling plants located in Chicago, Dallas and Los Angeles. In 1897, the product was already arriving in Canada, Mexico and Hawaii.

The company pioneered packaging design strategies in order to differentiate the product from the competition. The strategy was "*even in the dark or blindfolded*", the consumer should be able to identify the product by the differentiated design of its packaging. In 1918, the company moved to the command of Robert Woodruff, decisive for the worldwide consolidation of the brand and the leadership of the company in the sector. New marketing prospects were explored, venturing in communication (sponsorships for sports and cultural activities) and marketing (promotional packaging, exclusive contracts and refrigerator for product display). He also explored prices, and here we have a special chapter from the times of world conflicts. During World War II, the company promoted a special sales campaign for American fighters: the right to buy Coca-Cola anywhere for five cents (forty cents off the regular price of the product). In this phase, the company inaugurated 64 bottling units and reached in Europe. After the war, the business continued to accelerate, consolidating its presence in 100 countries by 1957.

How it all started: The chocolate kings[14]
The first records of chocolate bar date back to 1847, when Joseph Fry accidentally extracted cocoa butter; which was mixed with cocoa powder and sugar, forming a 'moldable paste'. With this invention, Joseph Fry opened the first known chocolate bar factory, named Fry and Sons. This new and emerging industrial segment benefited from three technological advances. In 1828 Van Houten invented a hydraulic press. In 1875, the partnership between a Swiss confectioner, Daniel Peter, and an entrepreneur, Henri Nestlé, culminated in the first milk chocolate bar. Lastly, in 1879 Rodolphe Lindt developed the conching technique.

This was the context of the emergence of some of the well-known brands in the chocolate industry. Nevertheless, war interests partially deviated the sectors' initial path, because governments and their Armed Forces were

---

[14] Based on Maxwell (1996), Coe and Coe (2000), O Mundo do Chocolate (2010), Mirrer (2012), and Reina (2015), as well as information retrieved from Mars website regarding 'The history of Mars', see at http://www.mars.com/global/about-mars/mars-pia/our-approach-to-business/story-of-mars.aspx.

attracted to chocolate as a strategic energy source due to its logistic facility. Companies had to face the challenge of creating a heat-resistant product that was also unattractive to troops, in order to postpone consumption for occasions with food restriction.

With this in mind, in 1937 the United States Armed Forces requested Hershey's to develop a product that was deliberately sensory unappealing: "*a bar weighing about four ounces, able to withstand high temperatures, presenting high food energy value, and tasting just a little better than a boiled potato*". The result was the development of the 'Ration D' (Figure), an extremely hard product, a bar consisting of six tablets of chocolate that would crumble with some effort and was resistant to temperatures of 120 °F (49 °C).

Mars's history also goes back to the battlefields. The company was founded in 1940, dedicated to producing a similar chocolate to the one found during the Spanish Civil War, sugar-coated pastilles, which provided greater heat resistance.

This innovation was employed in troop rations during World War II, from 1939 to 1945. Perhaps influenced by the relatively better flavor of the sugar-coated M&M's candies (Figure), the United States Armed Forces demanded from Hershey's a more palatable chocolate bar, with an only restriction regarding physical resistance to heat.

This moment led to the creation of a product that would put the company in history's hall of fame: 'Hershey's tropical chocolate', with sales that surpassed previous records in transactions with the United States Government. In 1939, daily productions reached values around 100,000 units. In 1945, with the introduction of 'Hershey's tropical chocolate', production was estimated as 24 million units per week. Throughout World War II, the estimated volume distributed to troops in combat had exceeded three billion units. It was an indication that new times were approaching in terms of industry orientation, with unfoldings regarding the role of processed food, concerning its representation and function in society.

Of course, the origin of the contemporary food industry is not restricted to war interests, having several business disconnected from this sad period of humanity. See **Box The kings of the meat.**

Anyway, as revised in Silva et al. (2018), the end of the conflicts culminated in the suspension of supply contracts with governments and their armies. As a consequence, the food industry was forced to search for new clients. The economic boom at that time stimulated mass consumption and an appetite for curiosity and convenience. And so, processed food entered into day-to-day family activities. In short, the food industry and its supply of canned processed food seemed to be in the right place at the right time (Silva et al. 2018). See **Box You got a 'SPAM'.**

However, what seemed to be the perfect convergence between supply and demand, represented the challenge of developing other bases for differentiating its product (Silva et al. 2018). It all started with the search for convenience and taste. Drivers that did not make any sense on the battlefield, but that on the consumer's table were basic requirements for the sector's competitiveness. It was the first time that the consumer was at the forefront of innovation drivers. Important remark other protagonists cannot be overlooked, such as the government through public policies regulating the sector (Law 2003), and the retail with its private quality policies, with an important development in the food system (Silva et al. 2018).[15]

The Meat kings[16]

At the end of the 19th century, the industry of refrigeration and related services (transportation and industrial equipment) originated from the dispute of two precursors of the refrigeration industry. Soon afterward, a concept was developed for domestic application, with the first refrigerator conceived in 1913. The Armor and Company was founded in Chicago/USA, making history as one of the precursors in the dissemination of the principles of production organization that emerged at the time: production lines in the slaughtering and processing of animal protein (beef and pork). Besides this production system, the company has also made history innovating in products, being the forerunner in the corned beef segment.

The credit that history grants to its founder, Philip Danforth Armor, is the concept of maximum use of the animal carcass. At this point in history, another emblematic character was beginning to gain strength. In 1853, Gustavus Franklin Swift, at the age of 14, started working in his family business, a meat merchant warehouse where cattle, sheep and pigs were raised in their own farm, in the American state of Colorado. Shortly thereafter, in 1855, with a US$ 20 loan, he opened a small butcher shop in the city of Cape Cod, Massachusetts. In 1869, he innovated in the beef market, starting to offer differentiated cuts of meat to his clients. In 1875, he moved to Chicago, and in 1878, in partnership with his brother, he founded a slaughterhouse, which was named Swift & Company.

The company was one of the first to invest in production verticalization, also investing in refrigerated transportation innovations. This sector culminated in the opening of a new company by the Swift group, dedicated to the service of wagons for live cattle transportation and refrigerated wagons, with

---

[15] Reardon et al. (2004) discuss the role of retail in driving organizational and institutional changes, as well as technological changes.

[16] More details invention could be found in Bruce-Wallace (1966) and Redação Super (1988).

blocks of ice inside, to transport animal carcasses. In 1877, the company successfully transported the first shipment of fresh meat in refrigerated wagons to the western USA. This initiative attracted the interest of the competitor Armour, who in 1883 founded Armor Refrigerator Line, a company specialized in the transportation of perishable foods, which became, in less than 7 years, one of the largest refrigerated truck fleets in the country (with a total of 12,000 vehicles in 1900).

The production of the fleet was verticalized in its factory of wagons. Armour also made history because of several controversies related to labor rights and accusations of product adulteration. The initiatives from Armour and Swift led to the concentration of the American refrigeration industry, which, throughout history, began to unfold through the strategy of mergers and acquisitions. In 1972, the two groups, Armour and Swift, merged to create the Swift-Armour company. After some other transactions, the brand was consolidated only as Swift, and, more recently, became part of the empire held by the Brazilian JBS-a character from the contemporary history of the sector.

You got a 'SPAM' by Hormel Foods Corporation[17]

Spam refers, in the digital world we currently live, to something 'undesirable'. But in fact, it is related to the short name for Spiced Ham and the company Hormel Foods Corporation.

Founded in 1891, in the city of Austin, Minnesota, by George A. Hormel who has made it to the list of the biggest fortunes in the United States, and his company got famous with one of its brands: SPAM.

Back in the 1970s, the British group Monty Python used the Spam brand in one of its parodies exactly as something we have no control over, due to its unwanted ubiquity (it was a criticism to the penetration of the company and its cheap product into the food bases) that goes beyond our control-a parody that came to influence, later, the terminology consecrated in the digital era (the 'e-mail spam').

---

[17] The original sketch Spam by Monty Python's Flying Circus can be seen in https://www.dailymotion.com/video/x2hwqlw

Table 1.1 Food science and technology knowledge development cycles and food processing industry paradigm shifts. (Source: Based on Silva et al. 2018)

Decades	Cycles	Food science and technology knowledge development	Industry reprogramming shifts	Drivers
1950	1st	Studies in quality control and water use reduction, which, in turn, were the basis for the development of new drying techniques and other innovations, such as osmotic dehydration, thermal microbial inactivation, refrigeration and freezing.	Convenience and Taste and Pleasure	New formats (frozen and refrigerated), textures and flavors, establishing the industry as a supplier of food solutions.
1970	2nd	Thermodynamics, rheology, biotechnology, emulsions, physical protection and image processing. In parallel to the study of the combination of different techniques as an alternative to milder processes (enhancing nutritional and sensorial characteristics of raw materials; and contributing to the taste of the product).		Product engineering (palatable formulations), as well as packaging and marketing strategies, seeking easy-to-taste products, despite nutritional aspects.
1980	3th	Ingredients replacement or removal, as well as the addition of nutrients and supplements, occurred within the industry.	Nutrition	The emergence of the categories 'better for you' and functional.
2000	4th	Wider scientific knowledge adopted from other disciplines.	Health and Wellness	Products that contribute to the consumer's health and wellness.

The process that followed culminated in the food industry transcending the pre-processing of agricultural inputs, to position itself as a supplier of food with added value in service and protection (Bruin and Jongen 2003).

In a short, after 150 years of products oriented towards accessibility (availability through extended shelf life) and food safety (microbiological control), the core business of the food industry reinvented itself. What went through three main paradigm shifts in the food processing industry, supported by four main Food Science and Technology knowledge development cycles from the mid-1950s, in the post-war period, until the beginning of the new millennium, See Table 1.1 (Silva et al. 2018). In the transition of each new paradigm, the search for better adherence to consumer demands.

From the focus in controlling seasonality and perishability of the raw material (first level), food industry turned in a sector dedicated to the processing of more elaborate products appeared, changing their *raison d'être*, from a mere product maker to providing service (convenience and flavor) and care (health, nutrition and wellness) (Bruin and Jongen 2003). Which in turn culminated in the development of new processing levels in the value chain involved in the transformation of the raw material into processed food (Silva et al. 2018). See ***Box The systemic game is the next big thing.***

All this reprogramming movement required breaking the process engineering paradigm over the product engineering, Fig. 1.2 (Silva et al. 2018). A shift that demanded a broader and interdisciplinary scientific knowledge, essential for the development of new processes and technologies (Aguilera 2006; Saguy et al. 2013) towards a new product conception and the understanding that several existing products have to be redesigned and new products will need to be created to satisfy the ongoing demand (Aguilera 2006: 1147).

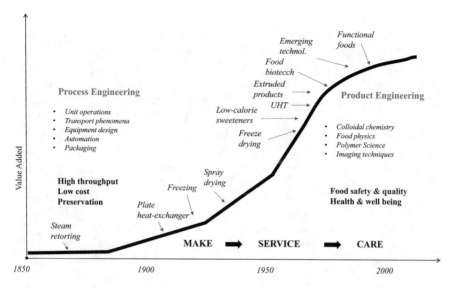

Fig. 1.2 Food industry breaking through: from process engineering paradigm over to product engineering. (Fonte: Aguilera 2006)

However, the contemporary challenge facing the industry is even more thought-provoking. As Silva et al. (2018) state, it is no longer a matter of desire for processing to conserve or to guarantee safety, seminal attributes of the industry, neither processing to potentiate taste and convenience, or contribute to the physical health of the consumer. The Box *Science on the plate* discusses this perspective.

The industry is challenged by new consuming trends towards social drivers related to food technology. Among their priorities, formulations that are more respectful to their physical health aligned to the health of the society and the environment in which the processing takes place. More than creating textures, the opportunity is also in preserving and valuing attributes derived from the raw material, in order to also preserve and ensure social and environmental welfare linked to the processing (Silva et al. 2018).

Starting from this exciting history and so everything to beginning, the main goal of this book is to deep in the contemporary context of the food industry, in order to discuss the challenges and opportunities for the industry-and the training of your professional. What follows next in the following chapters.

The Systemic Game Is the Next Big Thing

"The more the food industry moves to include service and care, the more effective its 'making' sector must become in order to keep economic margins. This implies very effective value chain management" (Bruin and Jongen 2003: 54). *"The major challenge is to deliver to the consumers branded products that satisfy their needs through the many stages of an optimized supply chain. This means a supply chain with short lead times in consumer demands, ..., with lower stocks and minimum off-spec products, and reduced waste/by-products"* (Bruin and Jongen 2003: 65).

Embedded in this discussion is the awakening of the sector to the consideration that organizational competitiveness permeates the boundaries of the company (Silva et al. 2018). The ability of the food industry to respond more effectively to business opportunities (in less lead time and costs) began to relate-more intensely-to the coordination of activities performed by other sectors along the value chain in which it is inserted (Batalha and Silva 2007; Besanko et al. 2012). What unfolds in joint planning with commercial partners, upstream and downstream, through the management of the different inherent flows, in particular, information, physical (product), financial (monetary) and negotiation flows. Assuming the systemic premise, for too long the convention was to illustrate the food processing value chain as linear. From farm to fork, in a logic flowing from left to right. A paradigm that has for long been questioned in virtue of the greater power of action and demand of the consumer, thus creating the agreement to treat the chain from right to left, from fork to farm. And more recently, linearity also begins to be questioned by a perspective since the linear economic model of production-consumption-disposal is reaching its limit. One way to address this problem is through an Economic Circular model, which associates economic growth with a continuous positive development cycle.

This paradigm is at the top of the international political agenda and it is expected to promote economic growth by creating new businesses and employment opportunities, reducing material costs, damping price volatility, improving supply security and, at the same time, reducing environmental impacts (Kalmykova et al. 2017). In its essence, circular economy concerns value aggregation in a resilient and sustainable way, based on the reintroduction and use of residues in the productive system/life cycle (agricultural or industrial), which come to be understood as by-products (Scheel 2016). It is time to consider that the first step in the transition from the linear logic to a circular economy is to expand the value proposition by identifying and capturing lost, not perceived, or not properly exploited values among the bonds involved throughout the productive system (agriculture-industrial-transformation-distribution-commercialization-use-disposal). A paradigm that, however, faces the creation of conditions that favor this transition, such as changes in the way society legislates, produces and consumes innovations, while also using nature as inspiration to respond to social and environmental needs (Kalmykova et al. 2017).

Science on the Plate

"When the first processed food was made available to consumers, they did not seem to mind eating products which were eerily indestructible; Twinkies, which seem to be virtually indestructible and possess an unnaturally long shelf life" (Wan 2012: 1). With the evolution, the sector has undergone, however, processes deliberately designed to make the product indestructible and even guarantee the homogeneity of the final product (blending strategies, masking variations in the quality of the raw material) began to be questioned.

From this perception, a new orientation of the industry gained relevance, leading to investments in process and product innovations. Among other examples, Nestlé announced investments to develop new flavor enhancement techniques that associate low temperature and high pressure.

Initial studies indicate that cooking food at high altitudes, where the pressure is lower, can make the flavor, aroma and color, more intense, as well as increase the potential to improving the nutritional quality of the food.

References

Aguilera, J. M. (2006). Perspective Seligman Lecture 2005. Food product engineering: Building the right structures. *Journal of the Science of Food and Agriculture, 86*, 1147–1155.

Batalha, M. O., & Silva, A.L. (2007). Gerenciamento de sistemas agroindustriais: Definições, especificidades e correntes metodológicas. In M.O. Batalha (coord.), *Gestão agroindustrial* (3rd ed.). São Paulo: Atlas.

Besanko, D., Dranove, D., Schaefer, S., & Shanley, M. (2012). *Economics of strategy* (6th ed.). Danvers, MA: John Wiley & Sons.

Bruce-Wallace, L. G. (1966). Biography James Harrison (1816–1893). In *Australian dictionary of biography* (Vol. 1). Melbourne: Melbourne University Press. Retrieved from http://adb.anu.edu.au/biography/harrison-james-2165.

Bruin, S., & Jongen, T. R. G. (2003). Food process engineering: The last 25 years and challenges ahead. *Comprehensive Reviews in Food Science and Food Safety, 2*, 42–80.

Calderón, L. A., Herrero, M., Laca, A., & Díaz, M. (2018). Environmental impact of a traditional cooked dish at four different manufacturing scales: From ready meal industry and catering company to traditional restaurant and homemade. *International Journal of Life Cycle Assessement, 23*(4), 811–823. Retrieved from https://link.springer.com/article/10.1007%2Fs11367-017-1326-7.

Coe, S. D., & Coe, M. D. (2000). *The true history of chocolate*. New York: Thames and Hudson.

Conway, C. (2012). *One billion hungry: Can we feed the world?* Ithaca, NY: Cornell University Press.

DEFRA. (2017). *Food statistics pocketbook 2016*. London: Department for Environment Food and Rural Affairs (DEFRA). Retrieved from https://assets.publishing.service.gov.uk/government/uploads/system/uploads/attachment_data/file/608426/foodpocketbook-2016report-rev-12apr17.pdf.

Euromonitor International, (2016). Global packaged food market to be worth US$2.2 trillion by 2021. Euromonitor International. Retrived from: https://blog.euromonitor.com/global-packaged-food-market-worth-us2-2-trillion-2021/.

FAO. (2018). Food loss and waste and the right to adequate food: making the connection. Rome. 48 pp. Licence: CC BY-NC-SA 3.0 IGO. Retrived from: https://www.fao.org/3/ca1397en/CA1397EN.pdf.

Floros, J. D., Newsome, R., Fisher, W., Barbosa-Canovas, G. V., Chen, H., Dunne, P., German, J. B., Hall, R. L., Heldman, D. R., Karwe, M. V., et al. (2010). Feeding the world today and tomorrow: The importance of food science and technology. An IFT Scientific Review. *Comprehensive Reviews in Food Science and Food Safety, 9*, 572–599.

IMAP. (2010). *Food and beverage industry global report 2010.* IMAP. Retrieved from http://www.proman.fi/sites/default/files/Food%20%26%20beverage%20global%20report%202010_0.pdf.

Kalmykova, Y., Sadagopan, M., & Rosado, L. (2017). Circular economy—From review of theories and practices to development of implementation tools. *Resources, Conservation & Recycling, 135*, 190–201. https://doi.org/10.1016/j.resconrec.2017.10.034.

Latour, B. (1984). *Les microbes: Guerre et paix.* Paris: Anne-Marie Métailié.

Latour, B. (1988). *The pasteurization of France.* Cambridge, MA: Harvard University Press.

Law, M. T. (2003). The origins of state pure food regulation. *The Journal of Economic History, 63*(4), 1103. Retrieved from http://www.jstor.org/stable/3132366.

Lazarides, H. (2012). Challenges and opportunities for the community of food sciences to contribute towards a society of healthier consumers and better world. *International Journal of Food Studies, 1*, 101.

Lopez, J. (2014). Changing role of business. *Creating Shared Value Forum, 2014.* Retrieved from https://www.youtube.com/watch?v=SOz5Qw1kc84.

Maxwell, K. (1996). A saga do chocolate. Folha de São Paulo. Retrieved from http://www1.folha.uol.com.br/fsp/1996/12/29/mais!/6.html#

Mirrer, L. (2012). *How World War II changed everything: Even our taste for candy.* New York: Huff Post. Retrieved from http://www.huffingtonpost.com/louise-mirrer/how-world-war-ii-changed-_b_2024730.html.

Nicolaou, A. (2018). Food industry giants struggle to keep up with changing tastes. *Financial Times..* Retrieved from https://www.ft.com/content/c30560e4-132c-11e8-8cb6-b9ccc4c4dbbb

Nunes, R., Silva, V.L., Kasemodel, M.G., Polizer, Y.J., Saes, M.S.M., & Trindade, C.S.F. (2019). A cross country analysis on the consumption of health and wellness food products. Working paper.

O Mundo do Chocolate. (2010). O chocolate e as Forças Armadas dos Estados Unidos. O Mundo do Chocolate. Retrieved from http://o-mundo-do-chocolate.blogspot.com.br/2010/05/o-chocolate-e-as-forcas-armadas-dos.html

Pollan, M. (2009). *In defense of food: An eater's manifesto* (1st ed.). New York: Penguin Books. 148 p.

Reardon, T., Timmer, P., & Berdegue, J. (2004). The rapid rise of supermarkets in developing countries: Induced organizational, institutional, and technological change in agrifood systems. *Electronic Journal of Agricultural and Development Economics, 1*(2), 168.

Redação Super. (1988). A geladeira. Super interessante. Retrieved from http://super.abril.com.br/comportamento/a-geladeira

Reina, A.S. (2015). A alimentação das tropas durante a guerra. Portal FEB. Retrieved from http://www.portalfeb.com.br/a-alimentacao-das-tropas-durante-a-guerra/

Saguy, I. S., Singh, R. P., Johnson, T., Fryer, P. J., & Sastry, S. K. (2013). Challenges facing food engineering. *Journal of Food Engineering, 119*, 332–342. https://doi.org/10.1016/j.jfoodeng.2013.05.031.

Scheel, C. (2016). Beyond sustainability. Transforming industrial zero-valued residues into increasing economic returns. *Journal of Cleaner Production, 131*, 376–386. https://doi.org/10.1016/j.jclepro.2016.05.018.

Segatto, C. (2010). Ervas medicinais: Os conselhos de Drauzio Varella. Revista Época. Retrieved from http://revistaepoca.globo.com/Revista/Epoca/0,,EMI162899-15230,00-ERVAS+MEDICINAIS+OS+CONSELHOS+DE+DRAUZIO+VARELLA.html

Silva, V. L., Kasemodel, M. G. C., Makishi, F., Souza, R. C., & Santos, V. M. (2019). Beauty is not only skin deep: Food ethics and consequences to the value chain. *Internext, 14*, 286–303. https://doi.org/10.18568/internext.v14i3.489.

Silva, V. L., & Sanjuán, N. (2019). Opening up the black box: A systematic literature review of life cycle assessment in alternative food processing technologies. *Journal of Food Engineering, 1*, 1–15. https://doi.org/10.1016/j.jfoodeng.2019.01.010.

Silva, V. L., Sereno, A. M., & Sobral, P. J. A. (2018). Food Industry and processing technology: On time to harmonize technology and social drivers. *Food Engineering Reviews, 10*, 1–13. https://doi.org/10.1007/s12393-017-9164-8.

Silveira, J. (2008, Nov). O perigo do chazinho. Folha de São Paulo. Caderno Saúde. Retrieved from http://www1.folha.uol.com.br/fsp/equilibrio/eq3010200805.htm

Singh, R.P. (2012). *Romancing with Food Engineering: A life-long second partner*. Annual Meeting of the Institute of Food Technologists, Las Vegas. Retrieved from https://www.youtube.com/watch?v=23eqfw2aaI8

Smil, V. (2001). *Feeding the world: A challenge for the 21st century*. Cambridge, MA: MIT Press.

The Center for Food Integrity (2018). A dangerous food disconnect. When consumer hold you responsible but don't trust you. The Center for Food Integrity. Retrieved from: https://www.foodintegrity.org/wp-content/uploads/2018/01/CFI_Research_8pg_010918_final_web_REV2-1.pdf

Varella, D. (2015). Reclama pro bispo. Folha de São Paulo, Caderno Ilustrada.

Wan, S. (2012). *Evolution in the processed foods industry: Exploring the impact of the health foods movement*. 98 f. Tese (Doutorado)—University of Southern California, LA, EUA.

Weaver, C. M., Dwyer, J., Fulgoni, V. L., III, King, J. C., Leveille, G. A., MacDonald, R. S., Ordovas, J., & Schnakenberg, D. (2014). Processed foods: Contributions to nutrition. *American Journal of Clinical Nutrition, 99*, 1525.

Chapter 2
The New Norm

"People buy foods for the benefits they provide, and how the foods fit into personal and societal values such as sustainability and environmental."

Weaver et al. (2014)

Abstract Social drivers in food technology. Reprogramming the industry, following a business sense. There is no escape from this route in which food ethics represents 'the new norm'. This movement opened a wide and diversified range of opportunities to the sector, derived from rethinking the technological bases (*status quo*) to favor consumer health, as well as the welfare of both society and environment in which the industry operates. To illuminate this new norm, in terms of theoretical approach and food market opportunity, is the main goal of this chapter.

Keywords Paradigm shifts · Food market · Food trends · Market intelligence · Food ethics · Shared value · Relationship · Sustainability · Environmental · Health and Wellness · Social drivers · Human resources training

2.1 Food Ethics

There has never been so much talk about ethics as these days. What brings us to the question: what is the ethics that so much is talked about? Among different definitions, it can be considered that ethics refers to the critical reflection on habits and customs, whether individual or of a group, for the benefit of the collective development and can encompass social and environmental aspects. Habits and customs that could be denominated by 'moral' in terms of the individual or society; and 'technological *status quo*' in terms of firms.

As revisited in Chap. 1, food industry-since its seminal conception, back in the 19th century, until the current days-has reinvented itself a few times in terms of its technological and scientific bases: from vegetable and animal raw material processer, with the main purpose of safety and conservation, to supplier of safe, practical and convenient food, ensuring sensory, nutritional and functional aspects (Aguilera 2006). Regardless of all the improvement of the sector, a clear evidence emerges: an important share of global consumers remains unsatisfied, revealing a more complex appetite towards social drivers on food processed products (Silva et al. 2018, 2019).

This movement opened a wide and diversified range of opportunities to the food system derived from the rethinking of the technological bases (*status quo*) in favor of consumer health, as well as the welfare of the society and environment in which the industry operates (Silva et al. 2018). See **Box** *Onion model*.

In fact, in addition to guaranteeing safety, accessibility, availability and variety of food and beverage products (Korthals 2001; Deblonde et al. 2007; De Tavernier 2012), the differentiation movement has been increasingly influenced by other aspects. Specifically, as discussed in Silva et al. (2019), contemporary consumers are increasingly demanding products that promote their own health and wellness, in alignment with animal and environmental welfare, as well as fair labor conditions in a global context related to the food value chain (Korthals 2001, Severo et al. 2018; Silva et al. 2018).

This concept is referred to as food ethics (Silva et al. 2019), referring to the meaning that in addition to assuring food security and safety, producers, processors, and retailers must consider ethical questions raised by production practices and conditions in the food value chain (Coff 2013, Deblonde et al. 2007). A new conception related to food processing comprising three main levels of consumer concerns (Brom 2000): *individual, groups of consumers*, and *citizens*. Starting from that, food safety is associated with food-related illnesses and so referring to individual and collective concerns, such as dietary restrictions or preferences related to personal or social beliefs and values, such as allergen-free. In turn, social concerns involves the promotion of sustainable development as well as the empowerment of local communities, as cleaner production and fair trade. While, environmental concerns could be associated with agroforestry systems, conscious use of natural resources, organic labeling, reduction of carbon footprint, management of water resources, encouragement of the principles of reduction, recycling, and reuse, among other concepts (Silva et al. 2019).

Onion model
It is important to emphasize that food ethics must be understood as one of the growth vectors of the contemporary industry. Ensuring both availability and regular access to safe food, also regarding nutritional terms, refer to an industry growth vector that cannot be overlooked as well. The other contemplates a market that, already meeting basic conditions, is driven by demands related to taste and convenience. At the extreme of this discussion, there is the longing for health and wellness at the micro (physical health) and macro (society and environment) levels, driven by consumers with better access to information and greater purchasing power.

These different drivers could compound an onion model in terms of differentiation strategies in food. At the center, the basic and fundamental food driver, *i.e.* integrity and availability in time of a safe food (shelf life). Which in turn could be involved by convenience, followed by taste and pleasure, and so nutrition and fitness, ended by health and wellness. A second remark is about not all layers of the onion need to compose the design nevertheless. Except its core referring to the basic precept of food. And finally, the thicknesses are not necessarily the same. It depends on different aspects such as socioeconomic conditions, access to information, supply, etc.

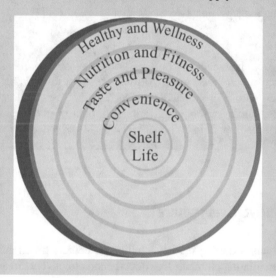

From that picture, the challenge of a new mindset of both the sector and its professional, i.e., the fascination and attraction of professionals over a process in the area do not make any more them capable of producing products that add value to the eyes of the consumer (Saguy et al. 2013).

Which brings the breakthrough based on the capability of imbricate the technological and scientific development in the social dynamics in which food processing is inserted (Silva et al. 2018, 2019). See **Boxes *Hunger for change***.

A context that highlights the dual social responsibility[1] of the food industry (Saguy et al. 2013): the health of the consumer (food safety) and the society in which it is inserted. Suggesting that, as important as process and product engineering (where differentials are built and ensured by processing and marketing technologies), it seems to be the relationship engineering established by the processing company in its value chain (Silva et al. 2018, 2019).

---

[1] Formulated by Bowen (1953), corporate social responsibility represents the purpose of aligning organizational activities with the values of the society in which organizations are inserted.

Hunger for change

Silva et al. (2019) provide a deep dive into the discussion of food ethics and opportunities in the food processor sector. According to them, from the three consumer concerns related to food ethics discussed above (*individual; groups of consumers; and citizens*), are expected three major shifts:

1. Food value chain coordination, in order to promote transparency and communication, bridging the gap between producers and consumers towards to solve problems of mutual distrust (Brom et al. 2004; Deblonde et al. 2007).
2. Technological and scientific approach, connecting the role of food in people's life (Beekman 2000; Brom 2000; Deblonde et al. 2007; Silva et al. 2018).
3. Competitive advantage background, going beyond aspects assured inside or within technological limits (Silva et al. 2018) towards outside aspects related to socio-environmental context regarding origination, processing, distribution and/or commercialization (Humphrey and Memedovic 2006; Henson and Humphrey 2009; Saes et al. 2014), Fig. 2.1.

2.2   Alternative Everywhere

There is a consensus among market intelligence centers highlighting food ethics aspects as differentiating elements that directly and effectively already impact daily food consumption.

This is, for example, what the British group Mintel reports show us. Among the elements identified as the most impacting on food and beverages consumption, it is found: "*Artificial: public enemy no.1*", "*Eco is the new reality*" and "*Based on a true story*" (Mintel 2016). In addition, another trend pointed out by the Mintel refers to "*alternatives everywhere*" (Mintel 2016). See **Box What's in my food?**

In other words, what was before considered alternatives now becomes conventional, intended to the general public, spreading the food intake over previously segregated groups. In this sense, the dissemination of the food ethics category, for all audiences, of all economies, makes history; once it is also expected or seen in developing countries, such as Brazil (Silva et al. 2019).

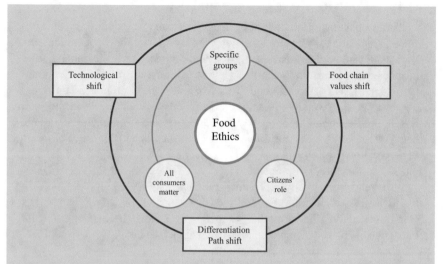

Fig. 2.1 Food ethics: product, society and environment. (Source: See at Silva et al. (2019) based on Brom (2000))

Ethical consumerism is no longer said to be restricted to developed economies. (…). As growing consumer concerns are increasingly reflected in the business plans of multinationals, consumers in emerging markets are becoming exposed to such issues as well (Euromonitor International 2014). This corroborates the understanding that, although differentials in food ethics should prove to be significantly more robust in developed economies, the potential of these strategies cannot be marginalized in developing regions (Euromonitor International 2014). Actually, they have the potential to drive product growth and stimulate the continuing growth of the category across all economies, since properly provided, including ensuring: transparency, certificates reputation, wide range and price repositioning (Euromonitor International 2014; Nunes et al. 2019). See the Box Are you elite now?

What's in My Food?

This question is getting more and more frequent among consumers, representing a trend that has been inspiring new businesses. Mobile apps that help us to translate food labels have spread all over the world since the early 2000s. The motivation is in how difficult it usually is to understand the information contained in the labels.

As discussed in Chap. 1, the relationship between consumer and food industry is still greatly hampered by the linguistic barrier in how companies communicate the composition of their products. One of the recurring provocations is that even though we have the ability to speak more than a language, "nutrition information on food labels is not one of them".

   This motivated the creation of businesses focused on empowering the consumer, elucidating what there is in their products and thus trying to guide their choices. Based on this idea, applications quickly began to contemplate other nuances of the decision,[2] such as specific issues about 'free from' food and even social drivers as origin and quality of life of rural producers. Different options are now available to the consumer, take a look at Fooducate[3] and EWG.[4] And in the preference dispute, new types of information are already being observed. Besides nutritional issues, certificates and degree of processing, you already observe among them concern for social aspects such as origin and sustainability of processing.

   A novelty that already represents a reality also in developing countries, like Brazil. As a reaction, the food industry gets into this game. Campbell's marketing action illustrates this repositioning: 'we believe people should know what's in our food: https://www.whatsinmyfood.com/'.

Are you elite now? Do you eat this healthy stuff? The Saladorama case[5]
Saladorama was born in the Brazilian city of Rio de Janeiro, with the proposal of delivering organic salads, made with fresh ingredients, with people from those communities participating in the processing and delivery. Employees, mostly women, go through a 4-month training process to learn everything from food care, such as cutting and sanitizing, to notions of business management, so that they are able to open their own business as well.

"We don't want them to be our employees. We want them to become entrepreneurs in the community." Four years later, in 2019, the format reached three other Brazilian cities (São Gonçalo/Rio de Janeiro, São Paulo city and Recife/Pernambuco).

From this picture, we could better characterize Saladorama company: a business model based on social impact, that is, a company that collaborates with the economic and social development of the region where it operates while generating profit to be reinvested in the operation itself. Customers are divided between residents of the communities (60%) and middle and high-class neighborhoods (40%).

---

[2] '10 Top Apps For Eating Healthy': https://www.forbes.com/sites/nextavenue/2013/08/27/an-app-a-day-keeps-the-doctor-away/#3bee4695543a

[3] Co-winner of the US Surgeon General's Healthy App Challenge for the healthy eating category in 2012. More details available in: https://wellkentucky.org/fooducate/ and http://www.fooducate.com/

[4] Since from 2014, https://www.ewg.org/foodscores/content/methodology

[5] Based in: Bergier (2015) and Fonseca (2019). More details available in: https://www.youtube.com/watch?v=RWT-p8XIROQ&+https%3A%2F%2Fwww.saladorama.com=

More than salad, consumers with higher purchasing power pay for the company's social value, subsidizing the sale of products at affordable prices to the population from low-income communities. In addition to this price strategy, from the final value of each salad, a percentage goes to the street artist responsible for creating the printing on the packaging. More than that, the images posted on the company's Facebook page have also audio descriptions so that visually impaired people can understand as well. According to the company, more than 11 thousand people have already been impacted by the delivery of 3 thousand salads, with average monthly net sales of 6.5 thousand Reais (1.6 thousand dollars), with a large part reinvested in the company itself. As a goal, more than profit, the dream of impacting, even more, the society through processing, and marketing healthy food.

Indeed, the relevance of health and wellness claim in processed food is outstanding in retail chains around the world, being food products with social claims no longer restricted to developed economies (Nunes et al. 2019; Silva et al. 2019). These drivers are deeply influencing the development of fast-moving consumer goods as well as aiming at the general public and, furthermore, appears on the shelves of retailer markets with strategic positioning in cost leadership. From these new consumption drivers, conventional products are losing market to differentiated options in terms of food ethics (Nicolaou 2018).

This fact is proven by the number of food and beverage products launched with an ethical claim, which strategy that gained strength over time (Silva et al. 2019). According to Mintel (2017), this has been a strategical target for many brands worldwide. In the period from 2010 to 2017, around 30,000 products were launched presenting ethical descriptions, certifications, and seals on its packaging as a manner to communicate socially and environmentally responsible practices (Mintel 2017).

A total of 28.082 products with an ethical claim were launched worldwide between 2010 and 2016. Regarding specifically food and beverage products, 7.441 were launched worldwide In 2016; against 1.362 in 2010 (Mintel 2017). Desserts and ice creams, fruit and vegetable, and bakery presented the most relevant growth over the period, presenting respectively an average growth rate of 61%, 56% and 49% per year, (Mintel 2017). Germany, the USA, and France together accounted for over 36% of all new products in the category. In turn, Brazil takes a prominent position and is inserted among big players in the food ethical market (Fig. 2.2.).

But how exactly is this happening? What do food retail gondolas show us? And what do the information on food labels reveal to us regarding social drivers on food processing? These questions take us to the next chapter of this book.

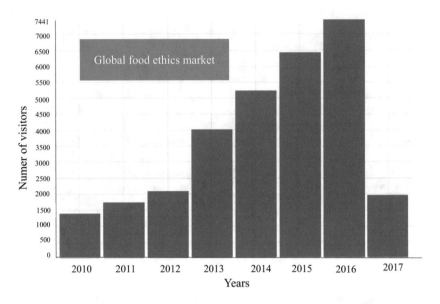

Fig. 2.2 At a glance: food ethics market around the world 2010/2017 (Source: Mintel (2017)

Country	2010	2011	2012	2013	2014	2015	2016	2017	Total Sample
USA	21.177	20.434	20.254	20.881	19.538	17.145	21.437	6.486	147.352
Germany	9.760	9.896	11.551	13.321	14.367	16.006	14.071	4.904	93.876
France	7.238	8.328	9.791	11.439	12.623	14.882	14.898	4.643	83.842
UK	8.433	10.693	12.415	11.352	9.635	12.521	12.188	4.243	81.480
India	4.852	6.829	8.481	14.421	14.606	11.979	11.829	3.602	76.599
China	5.875	8.218	9.263	8.916	11.675	11.326	11.825	3.798	70.896
Japan	8.236	7.860	8.369	9.269	11.086	10.218	9.694	3.080	67.812
Brazil	6.032	5.815	5.990	7.560	8.544	10.230	9.847	3.281	57.299
Spain	4.752	5.432	7.023	7.991	9.793	8.262	8.534	2.432	54.219
Canada	5.521	6.352	5.185	7.576	8.901	7.518	8.489	2.686	52.228
Italy	4.937	5.522	6.957	8.456	7.606	7.660	7.545	2.567	51.250
Mexico	5.180	4.160	3.591	5.028	6.516	6.029	6.277	2.084	38.865
S. Africa	3.503	3.744	4.896	5.444	5.789	4.433	5.285	1.695	34.789
Australia	5.403	3.811	4.219	3.860	3.680	4.888	3.961	1.178	31.000
Vietnam	2.975	3.640	3.860	5.228	4.538	4.063	4.440	1.218	29.962
Thailand	4.326	3.636	2.954	3.170	4.134	3.770	4.477	1.348	27.815
S. Korea	1.433	2.634	3.194	3.185	4.200	4.486	5.255	1.799	26.186
Poland	1.006	1.529	2.169	2.646	5.594	5.346	5.634	1.686	25.610
Netherlands	3.356	2.933	2.310	2.468	3.916	4.154	3.640	1.168	23.945
Argentina	2.382	3.017	3.076	2.953	3.496	3.634	3.211	1.008	22.777
Indonesia	1.027	560	794	2.073	3.687	5.085	5.509	1.631	20.366
Russia	2.371	1.402	1.892	2.317	2.897	3.664	3.254	1.064	18.861
Sweden	760	860	819	1.022	2.706	2.027	2.817	901	11.912
Total Sample	151.920	155.704	170.951	199.575	225.709	241.717	250.829	78.895	1.475.300

References

Aguilera, J.M. Perspective Seligman Lecture 2005 Food product engineering: building the right structures. Journal of the Science of Food and Agriculture, 86, 2006.

Beekman, V. (2000). You are what you eat: Meat, novel protein foods and consumptive freedom. *Journal of Agricultural and Environmental Ethics, 12*(2), 185–196.

Bergier, C. 2015. Saladorama, um delivery de saladas orgânicas feitas por moradoras de favelas: Para elas e para o asfalto. Draft. Retrieved from http://www.projetodraft.com/saladorama-um-delivery-de-saladas-organicas-feitas-por-moradoras-de-favelas-para-elas-e-para-o-asfalto/#sthash.Rv882jyM.dpuf

Bowen, H. R. B. (1953). *Social responsibilities of the businessman.* New York: Harper.

Brom, F. W. A. (2000). Food, consumer concerns, and trust: Food ethics for a globalizing Market. *Journal of Agricultural and Environmental Ethics, 12,* 127–139.

Coff, C. (2013). A semiotic approach to food and ethics in everyday life. *Journal of Agricultural and Environmental Ethics, 26*(4), 813–825.

De Tavernier, J. (2012). Food citizenship: Is there a duty for responsible consumption? *Journal of Agricultural and Environmental Ethics, 25,* 895–907.

Deblonde, M., de Graaff, R., & Brom, F. J. (2007). An ethical toolkit for food companies: Reflections on its use. *Journal of Agricultural and Environmental Ethics, 20,* 99–118. https://doi.org/10.1007/s10806-006-9019-4.

Euromonitor International. (2014). Corporate strategies in health and wellness: Part 1 focus developed markets. Euromonitor International.

Fonseca, A. (2019). O Saladorama enxugou a equipe, fechou filiais e adaptou as vendas para voltar a expandir, agora com foco. Draft. Retrieved from https://www.projetodraft.com/o-saladorama-enxugou-a-equipe-fechou-filiais-e-adaptou-as-vendas-para-voltar-a-expandir-agora-com-foco/

Henson, S., & Humphrey, J. (2009). The impacts of private food safety standards on the food chain and on public standard-setting processes, Joint FAO/WHO Food Standards Programme, Codex Alimentarius Commission, Third-second Session, Rome.

Humphrey, J., & Memedovic, O. (2006). Global value chains in the agrifood sector. Retrieved from http://tinyurl.com/y8qh4obd.

Korthals, M. (2001). Taking consumers seriously: Two concepts of consumer sovereignty. *Journal of Agricultural and Environmental Ethics, 14,* 201–2015.

Mintel. (2016). Global Food and Drink Trends 2016. Mintel. Retrieved from https://www.mintel.com/presscentre/food-and-drink/mintel-identifies-global-foodand-drink-trends-for-2016

Mintel. (2017). Global New Product Database. Mintel. Retrieved from http://brasil.mintel.com/gnpdbanco-de-datos-de-novos-productos

Nicolaou, A. (2018). Food industry giants struggle to keep up with changing tastes. *Financial Times.* Retrieved from https://www.ft.com/content/c30560e4-132c-11e8-8cb6-b9ccc4c4dbbb

Nunes, R. Silva, V.L. Kasemodel, M.G. Polizer, Y.J. Saes, M.S.M., & Trindade, C.S.F. (2019) A cross country analysis on the consumption of health and wellness food products. Working paper.

Saes, M. S. M., Silva, V. L., Nunes, R., & Gomes, T. M. (2014). Partnerships, learning, and adaptation: A cooperative founded by Japanese immigrants in the Amazon rainforest. *International Journal of Business and Social Science, 5,* 131–141.

Saguy, I. S., Singh, R. P., Johnson, T., Fryer, P. J., & Sastry, S. K. (2013). Challenges facing food engineering. *Journal of Food Engineering, 119,* 332–342. https://doi.org/10.1016/j.jfoodeng.2013.05.031.

Severo, E. A., Guimarães, J. C. F., & Dorion, E. C. H. (2018). Cleaner production, social responsibility and ecoinnovation: Generations' perception for a sustainable future. *Journal of Cleaner Production, 186,* 91–103.

Silva, V. L., Kasemodel, M. G. C., Makishi, F., Souza, R. C., & Santos, V. M. (2019). Beauty is not only skin deep: Food ethics and consequences to the value chain. *Internext, 14,* 286–303. https://doi.org/10.18568/internext.v14i3.489.

Silva, V. L., Sereno, A. M., & Sobral, P. J. A. (2018). Food Industry and processing technology: On time to harmonize technology and social drivers. *Food Engineering Reviews, 10*, 1–13. https://doi.org/10.1007/s12393-017-9164-8.

Weaver, C. M., Dwyer, J., Fulgoni, V. L., III, King, J. C., Leveille, G. A., MacDonald, R. S., Ordovas, J., & Schnakenberg, D. (2014). Processed foods: contributions to nutrition. *American Journal of Clinical Nutrition, 99*, 1525.

Chapter 3
In the Shopping Cart

"Food product has a lot of value, also referring to where (and how) it comes."

(Lopes 2014)

Abstract How effectively does this new norm is materialized today in supermarket chains? The main purpose of this chapter is to map and discuss the differentiation strategies practiced by the food industry, using an innovative proposal, based on an exploratory study regarding the decoding of information contained in food labels.

Keywords Food market · Food trends · Food products · Food label · Food packed · Label analysis · Reverse engineering · Food retail · Food value chain · Relationship · Origination · Sustainability · Environmental · Empowering producers

3.1   Rethinking Product and Process[1]

What is discussed here is the result of an empirical effort regarding baby food, chocolate, coffee, natural beverage, snacks and yogurt segments, conducted in two distinct moments, in France and Brazil during the year of 2016. And then updated in Brazil in 2019. The analysis was directed in terms of the countries, segments and retail chains considered. France represents the cradle of the contemporary food processing industry and Brazil takes a leading role among developing economies. On the other hand, coffee, chocolate, yogurt, baby food, snacks and natural drink

---

[1] Jose Lopez, Execute Vice President of Operations, Nestlé. Speech during 'Global Shared Value Creation Forum' 2014. See at: https://www.nestle.com/sites/default/files/assetlibrary/documents/media/events/csv-forum-2014/global-csv-forum-2014-summary.pdf.

© Springer Nature Switzerland AG 2020
V.-L. Silva, *Social Drivers In Food Technology*,
https://doi.org/10.1007/978-3-030-50374-1_3

portray emblematic cases of the global differentiation movement that has marked retail shelves. It is not intended to discuss market trends and niches. More than that, the main goal is to portray how processed food brands aimed at the general public position themselves in relation to the discussion of a new product conception. See Box *More than words.*

From retails shelves, it is prominent an effort by the industry to precisely review its technological bases through rethinking products and processes. As stated by Derek Yach, Ex-Director of Global Health and Agriculture Policy at PepsiCo, *"For decades food research centered on taste, not nutrition (and sustainability), so we are talking about pretty radical changes"*.[2]

Take for instance snack and baby food segments, in which formulations without preservatives or added sugar, as well as reduced sodium, fat and artificial colorings are a trend (Silva et al. 2019), in addition to rethinking processes in favor of sensory gains in texture, such as vegetables cooked separately and in smaller lots for instance. A small revolution started for new companies, which quickly made large companies reposition themselves as well. See the *Boxes Healthy finger foods* and *Momma's boy.*

More than words
A study derived from this empirical effort was published by Silva et al. (2019) by employing an original methodology based on Reverse Engineering. Beyond instructions for use, food labels contain information related to the technological steps and efforts present along the value chain. Examples of that are Seals and certificates (product, environmental and social), Marketing appeals (health, composition, processing and origin of raw material) and Social causes and philanthropic actions. This information was extracted from the analyzed labels, with the interest of discussing the companies' strategic positioning in terms of food ethics differentials and the impacts on the value chain.

Following that same direction, in natural beverages segment, it is quite clear a movement towards no added sugar, water or preservatives (Silva et al. 2019). The relevance of this new segment resulted in unfoldings in terms of reformulating existing lines of beverages, particularly nectars, such as the ones produced by the market leaders. See the *Box Juice box.*

Retail shelves also evidence food packaging strategies, differentiated in formats, colors and sizes with a social appeal associated with consumption. Once again, suggestive examples are found in baby food segments, constituting a clear positioning in response to critics concerning excessive consumption encouraged by family-size packaging. Additionally, baby food packaging also refers to the mother-child

---

[2] See at The Economist (2012).

relationship, as well as to the importance of value and sensorial stimuli, such as plastic packs that lower risks while stimulating child contact with food (Silva et al. 2019).

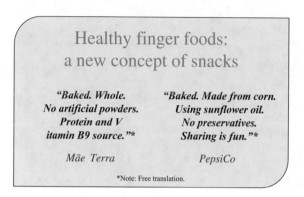

The examples above converge to a common denominator: the highlighted differentiating elements were constructed or assured within technological limits (product, process, marketing and packaging), without any direct unfolding in obtaining raw material (variety and origin), nor in the relationship with rural producers in favor of desired quality attributes (outside perspective).

Once again, an example is the differentiating strategy employed by the whole juice market, in which the consecrated branch is directed towards marketing differentials (inside perspective), instead of effective unfolding in the value chain (outside perspective).

One of the pioneer companies in this positioning is innocent drinks,[3] which revolutionized the market of natural beverages with the production of juices, smoothies and coconut water, creating a new market segment and starting new businesses abroad, including in Brazil. Nevertheless, this revolution was engineered through an inside perspective, within technological limits.

---

[3] Brand spelled using lower cases as is by the company.

> ## Juice box
>
> *"100% orange juice with fibers and vitamin C. The secret is caring."**
>
> Del Valle (Brazil)
>
> *"Whole orange juice. No added sugars, water or preservatives. Beverages made by young people tired of monotony."**
>
> do bem (Brazil)
>
> *"As innocent promise: We promise that innocent will always taste good and do you good. We promise that we'll never use concentrates, preservatives, stabilizers, or any weird stuff in our drinks. And we promise to brush our teeth twice a day.*
>
> innocent
>
> *Note: Free translation.

Joining on the bandwagon

In recent years, a trend has taken over the supermarket shelves: the clean label strategy. Product reformulation. Clean in terms of reducing ingredients. Seeking to recover homemade recipes. But especially presenting easy-to-understand consumer communication, lowering the existing language barrier.

A barrier that usually is translated by the colloquial names of the ingredients. In place of artificial thickeners, composition with apple. Instead of adding chemical conservatives, use natural ones. And instead of appearing on the label, for example, citric acid, write lemon.

"The dialogue around consumers' changing relationship with food is prevalent. (…). Consumers are telling all manufacturers that they want to recognize all of the ingredients in their food and that what they are consuming is made with the fewest ingredients possible". This emblematic phrase was said by the president of Hershey's when launching the repositioning of the company's brand and products.

It was in 2015, almost simultaneously with Nestlé. Since then, this movement has grown and spread (Watrous 2015). The global sales of clean-label products hit US$165 billion in 2015; been expected to reach $180 billion by 2020 (Bizzozero 2017).

This new strategy of communication/relationship between companies and their consumer is predicted to be in the late 2010s (Bizzozero 2017). About that, in a survey across Europe, North America and Asia-Pacific, 52% of respondents (out of a total of 1300 consumers) stated they would pay 10% more on a food or beverage product that contained ingredients they recognized and trusted.

Meanwhile, 18% would pay 75% or more extra; and 76% of respondents would be more likely to buy a product that contained ingredients they recognized and trusted (Bizzozero 2017). But there is an additional challenge: it is expected clean label becomes to be more holistic, encompassing the entire supply chain (Bizzozero 2017).

As discussed in the preceding chapters, one of the hot trends in food products is the true history behind the processing.

That means, in addition to product and process engineering, differentiation strategies should include social drivers, by rethinking the food industry relationships along its value chain, regarding the quality of relationships established by the industry with its stakeholders. A trend that has already become evident worldwide, including in developing countries, and that placing in checkmate position products consecrated by history (Nicolaou 2018).

Although all these examples are very suggestive, in terms of the sector's efforts to review its technological bases and also of communication with its consumer (theoretical aspects introduced in the previous chapter), the retails shelves reveal that this is one of the paths followed.

In complement, there is the strategy to go beyond technological limits, encompassing differentials built along the value chain. Another theoretical nuance discussed previously, next illustrated with real examples from the shelves.

3.2   Rethinking Relationships in the Value Chain

Let's to look out side the box. Looking for a different way to the new. In fact, retail shelves indications go beyond claims assured inside (product, process, marketing and packaging). In the base of this other differentiation strategy, retail shelves reveal food products with claims built along the value chain, referring to an outside perspective (initiatives that promote equally healthy relationships with society and the environment in which the industry operates).

In terms of the outside perspective of innovation, the organizational decision of brand differentiation shows three main axes, Fig. 3.1 (Silva et al. 2019).

The first axis refers to welfare programs, in terms of social or environmental initiatives. The case of innocent drinks is once again illustrative. The company seeks to differentiate its brand financing social and environmental projects (assured outside). They invested in organizational initiatives in favor of a gender equality policy.[4] In addition, at least 10% of innocent drinks annual profits are donated to charity, funding projects under two main goals: stop children dying from hunger

---

[4] https://www.innocentdrinks.co.uk/us/gender-pay-gap-report-2018.

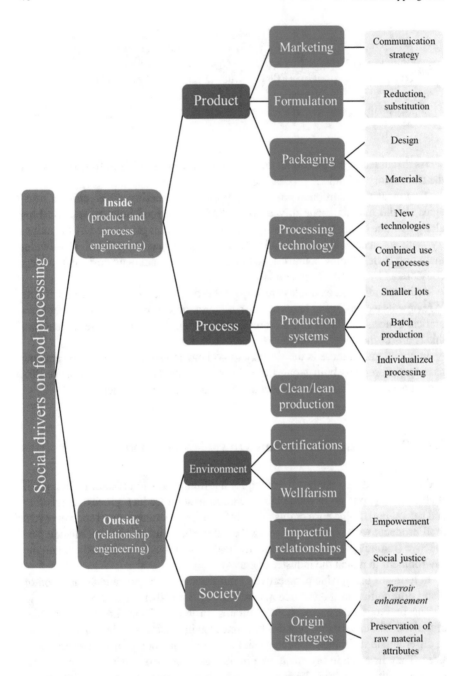

Fig. 3.1 Social drivers in food technology in a inside and outside decision map. (Source: Based on Silva et al. 2019)

and help the poorest families feed themselves.[5] Moreover, the company supports environmental initiatives such as a green revolution in plastic bottles. See **Box What about the rest?**

What About the Rest?

Our mission "getting healthy drinks to people to help them live well and die old. At the moment, plastic bottles are the most sustainable way of doing that. We're on the case for making sure that plastic is treated in the right way, so it has much less impact on the environment".

"Our new smoothie bottle is our most sustainable bottle yet. It's made up of 50% recycled and 15% plant plastic. We use a by-product of sugar cane called molasses in the plant bit. We're working on a way to use no oil-based plastic at all in the future, making our bottles 100% renewable. We're well on our way to making that happen by 2022. We'll keep you posted.

This kind of initiative, related to brand strategies guided by environmental healthiness, is also widespread in Brazil. Among various other examples, we highlight Danone's initiative. In Brazil, since 2011 the company innovated by promoting the use of green polyethylene with the I'm green™ certification. In the same year, the company started to invest in a circular economy initiative with pickers in more than 60 cities in the Southeastern Brazil region.[6] While in France, Danone was ahead of the packaging reduction movement (secondary packaging).

The second axis refers specifically to the relationship of the food processing industry with its partners along the value chain, which involves public and private certificates on human and social development (Silva et al. 2019). Examples come from Fair Trade, UTZ and Rainforest Alliance certificates, as well as agriculture and fishing sustainable.

Although still representing a positioning modestly employed in the Brazilian market, certain initiatives may not be overlooked in terms of the second axis. Among these, it is highlighted Family Farming ('A*qui tem agricultura familiar*') and EcoSocial All Fair certificates. The previous concept identifies products originating from family farms in Brazil and aims to promote the strengthening of the segment's social identity with consumers and the population. Launched in 2008, the certificate has benefited more than 200 million producers, spreading over products sold in Brazil.[7] On the other hand, the EcoSocial All Fair is a certificate from the Brazilian Institute of Development (IBD) which is applicable exclusively to organic products

---

[5] https://www.innocentdrinks.co.uk/us/sustainability/protected-futures/innocent-foundation.

[6] http://www.letstalkpackaging.com.br/noticia/danone-e-o-processo-de-inovacao.

[7] https://www.embalagemmarca.com.br/2011/08/danone-adota-polietileno-verde-da-braskem-nas-embalagens-de-activia-e-danoninho/.

and processes. The label can be obtained by companies, properties or producers that indicate a process of human, social and environmental development fostered by commercial relations based on the principals of Fair Trade. It was observed, for instance, amongst products from Native, a Brazilian company producer of a wide range of organic products, including sugars, snacks, coffees and chocolates.

Another trend is bio food products. **See Box** *Companies go blue*. The case of Danone and Nestlé, and, following international movements, *B Corp* certificate, based on healthy social relationships along the value chain. From Brazilian shelves, Unilever and its new brand 'Mãe Terra' is helping to spread this certificate in the processed food market, particularly in the snacks segment. More than a healthy snack to the consumer. A healthy product for both society and the environment.

A good example comes from Ethiquable, a brand of coffee chocolates, teas, among other products, which is committed to the practice of Fair Trade and AVSF (*Agronomiques & Veterinaires Sans Frontières*). Acting in twenty developing countries, the company supports the agricultural production of local producers. Another brand that stands out in terms of the relationship established with its partners is Tonny's, and your commitment as a company structured in a 100% slave-free cocoa supply chain. Unfortunately, a practice still recurring today in cocoa farms in Africa (Silva et al. 2018). This demand has been stimulating the birth of new brands such as Tonny's as well influencing reacting from the big players. See *Box Chocolate made from relationships*.

Companies Go Blue

Among other initiatives, the B Corps Certificate stands out. Created in 2006, they define themselves as "*a community of leaders and drive a global movement of people using business as a force for good*". Pursuing this goal, the label attests highest standards of social and environmental performance, public transparency, and legal accountability, in terms of a better balancing between profit and purpose. The certified companies use profits and growth to generate a positive impact on their employees, communities, and the environment. In 2019, more than 3200 companies, from 71 countries and one unifying goal, "*people using business as a force for good*". More information: https://bcorporation.net/

Finally, it is also spotted on the retail shelves a trend towards flavor and texture preservation strategies, constituting the third axis, which aims the maintenance of natural characteristics associated with the raw material (Silva et al. 2019).

This strategy is frequently linked to product origin, in which the quality of the final product is determined by attributes specific to the raw material origin, a contemporary vision may also refer to animal welfare issues. The world-renowned French standard of terroir valorization is an example. In French retail shelves, movement in this direction is pronounced even in the fast-moving consumer goods sector. This is particularly the case of the whole milk yogurt segment with no added sugars, preservatives or artificial ingredients, as suggested by label analysis.

Concerning this differentiation base, origin, form of confinement, animal breed and other factors interfere decisively in the quality of the final product. A differentiated raw material is necessarily obtained through particularly coordinated supply sources, in terms of relationship with rural producers.

Chocolate Made from Healthy Relationships

"Its about what I eat. That's what I stand for. Convenience (Ready to eat). More and more Health. More justice. Ethiquable supports NGO AVSF. The real 'light product' is the one without any pesticide-based treatment." *Ethiquable*

AVSF (Agronomes et Vétérinaires Sans Frontières) is a non-profit association that works for international solidarity and that has been engaged in supporting smallholder farming since 1977.

Source: Ethiquable packed photo from the Private collection of the author. About AVSF, see at: https://www.avsf.org/en/mission

https://tonyschocolonely.com/storage/configurations/tonyschocolonelycom.us/files/jaarfairslag/2017-2018/tonyjfs_201718_complete_eng.pdfmission;

https://www.raconteur.net/business-innovation/child-labour-cocoa-production
Source: https://www.barry-callebaut.com/sustainability

It is noteworthy that this particular movement is still considered a niche in Brazil, restricted to *premium* products that are commercialized in specific channels. However, a closer look evidences the start of a movement of dissemination of brands known to be differentiated, suggesting the start of a dispute for space with conventional products directed to the general public. One of that also comes from chocolate segment and the AMMA chocolate, a pun on the word love.

In 2002, its founders began the organic planting of cocoa on the family farm in southern Bahia near Itabuna, Ilhéus and Itacaré. In 2005, they participated at the Salon du Chocolat in Paris, presenting their organic cocoa to European chocolate makers, who thereafter began to use this noble raw material. Finally, in 2007, after a partnership with Frederick Schilling, founder of Dagoba Organic Chocolate in the USA, the dream came true, and AMMA Chocolates was founded. Their cocoa beans are originated from a demarcated region, a specific terroir highlighted in packaging, located in the Atlantic Forest (as shown in the **Box** **Loving chocolate**, 14° 19'S * 39° 15'W), free from health-damaging synthetic chemical fertilizers, and free from genetically modified organisms.

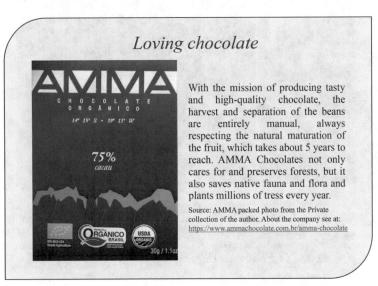

Loving chocolate

With the mission of producing tasty and high-quality chocolate, the harvest and separation of the beans are entirely manual, always respecting the natural maturation of the fruit, which takes about 5 years to reach. AMMA Chocolates not only cares for and preserves forests, but it also saves native fauna and flora and plants millions of tress every year.

Source: AMMA packed photo from the Private collection of the author. About the company see at: https://www.ammachocolate.com.br/amma-chocolate

3.2.1   Enjoy Your Food!

In short, we can conclude this Chapter highlighting the movement of brand differentiation structured in claims conducted inside (product and process engineering) and outside (relationship engineering) of the organizational boundaries. But why?

What else can we extract from this organizational decision?

What are the managerial implications of this decision easily understood as more expensive, especially for considering competencies not belonging to the core business of the company, i.e., processing technology and food conservation?

The center point of this questioning seems to be the need to break the main tendency in the analyses regarding the differentiation strategies practiced in the area of food science and technology.

Technology-if not supported by other factors-tends to be proven fragile in the organizational search for competitive differentials that sustains itself over time.

This is the main goal of interest of the next chapter.

References

Bizzozero. (2017). *75% of consumers will pay extra for clean label ingredients. Food beverage insider.* Retrieved from https://www.foodbeverageinsider.com/market-trends-analysis/75-consumers-will-pay-extra-clean-label-ingredients.

Lopez, J. (2014) *Changing role of business. Creating Shared Value Forum 2014.* Retrieved from https://www.youtube.com/watch?v=SOz5Qw1kc84.

Nicolaou, A. (2018). *Food industry giants struggle to keep up with changing tastes. Financial Times.* Retrieved from https://www.ft.com/content/c30560e4-132c-11e8-8cb6-b9ccc4c4dbbb.

Silva, V. L., Kasemodel, M. G. C., Makishi, F., Souza, R. C., & Santos, V. M. (2019). Beauty is not only skin deep: Food ethics and consequences to the value chain. *The Interne, 14*, 286–303. https://doi.org/10.18568/internext.v14i3.489.

Silva, V. L., Sereno, A. M., & Sobral, P. J. A. (2018). Food Industry and processing technology: On time to harmonize technology and social drivers. *Food Engineering Reviews, 10*, 1–13. https://doi.org/10.1007/s12393-017-9164-8.

Watrous, M. (2015). *Hershey switching to simpler ingredients. Baking Business.* Retrieved from https://www.bakingbusiness.com/articles/28207-hershey-switching-to-simpler-ingredients.

Chapter 4
Stay or Leave

"Sustainable development is not a virtue or a good deed; it is business sense."

Riboud (2013)

Franck Riboud, Diretor Executivo (CEO) Danone. Available at https://downtoearth.danone.com/2014/04/29/danones-2013-sustainability-report-is-available

Abstract Supermarket chains signal a movement of brand differentiation structured in actions conducted inside (product and process engineering) and outside (relationship engineering) of firm boundaries. Nevertheless, why does this happened? What factors can be attributed to this economic decision of 'leaving' its know how that may include competencies that are not a part of the core business of food companies and that seems to be more costly? At the heart of this questioning seems to be the need to break the main analysis bias concerning the strategy of differentiation practiced in the field of food science and technology. Technology-if not backed up by other factors-tends to prove itself fragile in the organizational longing for competitive differentials that are sustained over time. This is the main subject of interest in this chapter.

Keywords Food products · Food market · Food packed · Food innovation · Product and process innovation · Sustainable competitive advantage · Shared value · Food value chain · Relationship governance · Human resources training

4.1 A Business Sense

Important to open this Chapter rescuing one of the economic fundamentals. Technology adds value by allocating labor and capital in the processing (industrial or handcraft, under a continuous production system or in batch production), transforming the raw material from animal or vegetable into a processed product that carries benefits such as safety, convenience and practicality in preparation and consumption, including aspects related to shelf life (possibility of storage ensuring food integrity and stability). For example, cheese and dairy products, fruit jelly, roasted and ground green coffee illustrate value-added products.

From this perspective, technology also allows adding value in the form of health-differentiated products, such as functional foods, reformulated (baby food and snacks are again very illustrative) or even special foods (free from allergenic ingredients).[1]

In the field of food science and technology, including food engineering, this is the mindset regarding differentiation strategy. Under this path, the value aggregation paradigm can be associated with an inside perspective, where differentials are built (assured) by product and process engineering, encompassing product and packaging designs, as well as marketing strategies, such as brand licensing. This dynamic originated the four cycles of investments in food science and technology that have been seen since the conception of the contemporary industry, in the nineteenth century.[2]

An interesting example comes from Brazil related to both Greek type yogurt and differentiated chocolates as discussed next. Brazilian consumers have been driven to the shelves of dairy products in a search for yogurts classified as Greek yogurt. In contrast, however, with the original conception of the product (in Europe, Greece), the product standard disseminated in Brazil presents altered composition that allows ensuring, through processing (inside), the texture (sensory) and quantity of protein (nutritional).

Moreover, differentiated chocolate has been another significant consumption trend in Brazil. Led by this trend, the widespread practice among the major chocolate processors in Brazil, however, refers to innovations that are assured inside, restricted for example to product formulation (such as the dark variety), or packaging and marketing strategies. See ***Box As a gift***.

The ***Box Perspective inside*** shows the characteristic composition of the Greek yogurt in Brazil. Launched in 2012 by Vigor, only a few months later each one of the other giants in the sector already owned their respective brands of Greek type yogurt.

Two years later, in 2014, the product already represented 10% of the Brazilian yogurt market, constituting the portfolio of the main organizations in the country, including small-scale firms. In turn, another significant consumption trend in Brazil has been the search for chocolates announced as premium. Getting a free ride in this

---

[1] The technological perspective of value addition (inside) is also usually related by the processing company in terms of the superior quality on raw material used in the processing (denoting superior quality to the final product) or the tradition of the processing company in the market (unfolding in a greater confidence associated with the brand).

[2] See Chap. 1.

As a gift[3]

The emphasis on marketing and packaging strategies as the main value-adding strategy in chocolate is potentialized during Easter time.[4] In 2013, Kopenhagen, one of the leaders of the Brazilian premium chocolate market, released a limited edition (Egg Collection) of Eastern eggs presenting packaging studded with Swarovski crystals to celebrate its 85 years. Following the same trend, all the major companies operating in the Brazilian industry began to sell conventional products (chocolate) in packaging with differentiated design.

Another trend is the brand licensing, a marketing strategy specially focused on children's market, in which thematic toys and accessories are linked to the product. Beginning in the first decade of the 2000s, this marketing strategy quickly spread among all companies in the sector. Throughout this short trajectory, however, an important change has occurred: the toys and accessories are no longer inside the eggs, they are now outside, involving the product. As made by Lacta (Mondelez Corporation) in 2018: Jake and Finn, protagonists of the Adventure Time cartoon, are the packaging for mini Easter eggs. Innovation that has allowed companies to control one of the biggest problems affecting the Easter egg business: risks of physical damage during distribution logistics, shown in the images.

movement, as a general rule, the practice disseminated among the major chocolate processors in Brazil, however, refers to innovations guaranteed inside, restricted to product formulation (for the dark variety) or even to packaging and marketing strategies.

A similar movement was observed on chocolate in the Brazilian market, in regard to the dissemination and copy of the quality standards established by the first entrants. The dark chocolate variety arrived in the country in 2007. Among the pioneers were Garoto (Talento Intense™, 55% cocoa), and Hershey's (Special Dark™ 60%). Soon after, dark chocolates spread throughout the country, with the main companies of the sector owning their respective dark chocolate brands, with higher cocoa content. See *Box Perspective inside*. The category represented in 2011 the third most consumed type of chocolate in Brazil, representing an increase of 31% in sales value (Sá 2012).

---

[3] Based on Portugal (2013), Food Service News (2014), and Rogê (2014).
[4] Longest period of sales for the sector in Brazil. It would be 100 million eggs per year (Rogê 2014). In 2014, the production was 20.2 thousand tons, ABICAB (Brazilian Association for the Industry of Chocolate, Cocoa, Peanuts, Candies and Derivates), available in Food Service News (2014).

However, differentiators based on the inside perspective must be protected by means of patents or other entry barriers, such as required capital and brand value. Otherwise, inevitably, the differential is lost. See the **Box** *Kept under lock and key*.

This is exactly what has been observed for dairy products and chocolates in Brazil, in terms of the dissemination of Greek type yogurt and differentiated chocolates.

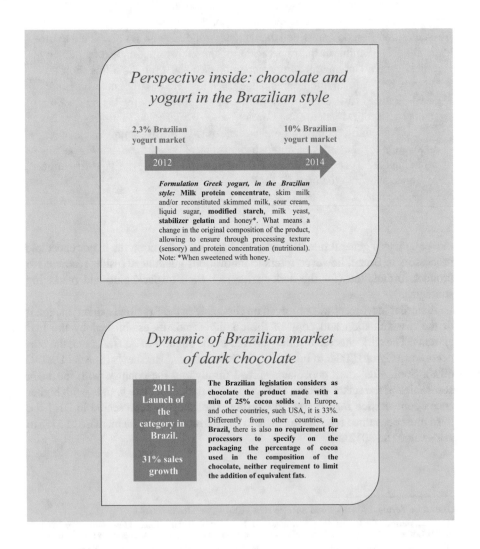

Kept under lock and key

Fundamental to register technological effort protected by patent. Chocolate case with collagen. And so many other very important innovations, revolutionizing the market and sector through disruptive innovations. A consecrated effort that denotes one of the main competences of the research and development area.

In the case of chocolate, a product with an empty sugar molecule, ensuring sweetness with a significant reduction in energy and sugar. Natural dyes. Fat reduction. All with disruptive technologies. Seductive and important subject. But we try to go further here. Bringing a look of innovation in R&D that starts in the field and has in its quality of relationship with stakeholders another fundamental ingredient of the formula protected under seven keys.

Take for instance the case of Milkybar, world first Nestle product with innovation to reduce sugar by 30%, a scientific patented breakthrough in which Nestlé changed the structure of sugar using only natural ingredients. "They created an aerated, porous sugar that dissolves more quickly in the mouth. This allows someone to perceive the same sweetness as before while consuming less sugar".

According to the company, Milkybar is one of Nestlé's most iconic chocolate brands. Launched in the UK in 1936, it is a popular choice for parents in the UK for their children thanks to its mild and creamy taste and high milk content. In 2007, the brand moved to all-natural ingredients. In 2017, milk became the No. 1 ingredient in the recipe. And in early 2018, *Milkybar Wowsomes* comes e with 30% less sugar. This product inovation is in line with Nestlé's purpose of enhancing quality of life and contributing to a healthier future. The company started its sugar reduction journey in 2000. It made a first public commitment to reduce sugars in a range of products by 10% between 2014 and 2016. Nestlé has pledged to further reduce the sugars in its products by a further 5% on average as part of a range of 2020 commitments. More information, see at:

https://www.nestle.com/media/news/nestle-milkybar-wowsomes-innovation-sugar-reduction

Launched in 2012 by Vigor, only a few months later each one of the other giants in the sector already owned their respective brands of Greek type yogurt. Two years later, in 2014, the product already represented 10% of the Brazilian yogurt market, constituting the portfolio of the main organizations in the country, including small-scale firms.

For the chocolates considered as differentiated, a similar movement was observed on the Brazilian market, in regard to the dissemination and copy of the standards established by the first entrants. The dark chocolate variety arrived in the country in 2007. In the search for competitive differentials that endures over time, the outside perspective emerges as a powerful alternative. At the heart of the differentiation is the fine relationship established between the quality of the processed food and the origination of the raw material, which may include the bases of the relationship with rural producers (Lopez 2014).

This association complies with the desire for a sustainable competitive advantage exactly because it represents a differentiating element that is not replaceable and difficult to copy, without costs and in the short term by the competition, for its unique character of the relationship in evidence (Barney 1991). Which goes through, as discussed in Silva et al. (2019), a true story behind processing (Mintel 2016), related to the socio-environmental context regarding origination, processing, distribution and/or commercialization (Humphrey and Memedovic 2006; Henson and Humphrey 2009; Saes and Silveira 2014).

The next section brings us two cases specially chosen seeking to enrich the discussion of the business sense behind the decision to differentiate from an outside perspective, i.e., from attributes developed and maintained along the value chain in which the food is inserted.

4.2 How About a Coffee?[5]

For more than a decade the coffee market has been reinventing itself with a new way of selling and appreciating its product. Coffee capsules also referred to as single-cup, represent a hot segment with significant worldwide growth, harmonizing quality control and standardization providing convenience and practicality for consumption.

This was Nestlé's first inspiration for the conception of what came to be the worldwide success of Nespresso.[6]

> The Nespresso story began with a simple idea: enable anyone to create the perfect cup of espresso coffee-just like a skilled barista
>   (Nestlé 2008)[7]

The invention dates back to 1976 by a collaborator of the Swiss company. But it was only after a decade, in 1986, that the new concept gained projection with Jean-Paul Gaillard, Nespresso director at the time, leading the way from 1988 to 1997.[8]

---

[5] Title inspired in The Brand Gym (2012). https://thebrandgym.com/espresso-wars-starbucks-takes-on-nespresso/.

[6] *"Single-cup* (shot) is the future. It is a coffee subjected to minimum errors, a drink with regularity" (Rosa 2013). In the 90s, this purpose guided the coffee roasting segment, but at the time heads turned to the sachet. Revolutionary at the time, the Italian Illy led the movement. Along the way, however, the glamor associated with Swiss capsule Nestlé and the possibility of having this refinement at home (a marketing trump card that illy did not see, focusing on its strategy in bars and restaurants) stole the scene. The sachet continues to exist and to be present in the sector. But the main target of the sector is the single-cup coffee in capsules.

[7] "Nespresso system: machine, coffee and service interconnected" (Bruegger, Nespresso Americas Director, Mundo S/A 2013). Ingenious system that controls temperature, quantity, speed and water pressure, which emphasizes the aroma, texture and flavor of the grains (Mundo S/A 2013).

[8] Two decades later, the Swiss company launched its second brand, Dolce Gusto, in 2006.

In three decades, the coffee capsules segment has been consolidated and is considered, by Howard Schultz, CEO of Starbucks, the fastest growing business in the global coffee industry (Lucas and Simonian 2012). Illustrating this assertion, facts show that from the US$ 60 billion dollars flow of the category, half comes from the commercialization of roasted coffee, while the other half comes from instant coffee and capsules. The capsule segment alone has a margin of 30%, three times higher than the other types of coffee, presenting prices five times higher than the kilogram practiced by roasted coffee brands (Isto É Dinheiro 2013). Representativeness that culminated in projections of US$ 12.58 billion in 2015, 57% higher than the US$ 8.03 billion observed in 2012, for the coffee capsule business (Agência Estado 2013).

In Brazil, the same worldwide trend is observed. According to Euromonitor International, the number of machines for the domestic preparation of coffee increased by 46% in Brazil from 2011 to 2012, jumping from 100 thousand to 146 thousand units. In turn, research conducted by ACNielsen Consulting indicates that capsule consumption increased by 46.5% from 2012 to 2013. In value, the increase represented 36.5% in the same period. By 2017, projections show a growth of 18% per year in the segment. This tendency is sustained by the incipience stage of Brazil in contrast to other markets already well consolidated. Only 0.6% of households in the country consume this product; being only 2% of the total coffee represented by capsules, versus 17% and 30% in the USA and in some European countries, such as France and Switzerland, respectively (Almeida 2013; Exame 2014a, b).

Behind these numbers is the performance of Nestlé company, which has built, along the three decades of history, a worldwide noticeable spot in the single cups market.[9] The Swiss company controls 40% of the world market for coffee capsules and the Nespresso[10] brand represents one of its most important businesses: providing double-digit annual growth since 2000; representing 4% of Nestlé's global sales in 2011; US$ 5 billion revenue in 2013 (Mundo S/A 2013) and a predicted US$ 12.58 billion in 2015 (Café Point 2013; Isto É Dinheiro 2013; Mundo S/A 2013).

Nestlé ruled alone the world capsule coffee segment for almost 25 years. But the expressive representativeness of this new format of business and product stimulated

---

[9] Nespresso is present in 60 countries. More than 60% of Nespresso coffee originates in Brazil, constituting itself as the main supplier of the brand (Mundo S/A 2013). The Brazilian market also stands out in Nespresso's business for receiving the facilities of the first capsule factory outside Europe, where three other units coexist. The new unit is located in the city of Montes Claros, Minas Gerais state, with operations starting in the second half of 2015 (Blecher 2014).

[10] Nespresso brand is known as the Louis Vuitton from coffee. Nestle's Nespresso coffee machine was recognized by Deutsche Bank analyst Jamie Isenwater as "the closest thing to a luxury brand in the fast-moving consumer goods segment." It operates on a 'closed' machine-capsule system that aims to prevent the use of competing coffee. The business model results in a higher density of sales in its stores than in Louis Vuitton, says Isenwater (Lucas and Simonian 2012).

investments for the purpose of copying the Nespresso System, which started questioning judicially the technological domain held by Nestlé.[11]

Innovation approved. Innovation (to be) copied

(Mundo S/A 2013).

Between 2011 and 2012, Nespresso's first patents began to fall into the public domain. At that time, new entrants positioned themselves in the market, investing in selling capsules compatible with the Nespresso machines (generic) and also independent systems to extract the drink.

The dispute went to court in different countries, like France, Germany, Belgium, Spain and Netherlands, in judicial processes involving three main companies: the Dutch company D.E. Master Blenders 1753,[12] the Swiss Ethical Coffee Company[13] and the German Dualit.[14] The first two companies were the pioneers in selling coffee capsules compatible with the Nespresso machines. Duality, however, marked the first judicial decision against Nestlé, in which it was legitimized the argument that the consumers, when buying the machine, have the rights to operate it in the way it best suits them, including using of generic capsules (Paladar 2013).[15]

---

[11] From the first registration in 1996 to the first decade of the 2000s, Nespresso held more than 1700 patents, with different expiration dates and contemplating different aspects related to capsules and the operating system, including design and packaging technologies as well as extraction of the beverage. According to the company, new patents are constantly being created: "Since the system itself has been the target of continuous innovations, there are many different patents related to these capsules, for which the claim of compatibility must be evaluated" (Almeida 2013). In this respect, it is worth considering that the patents are valid for 20 years from the registration application date. But there is a strategy to extend that time: the utility patent, whereby the inventor adds changes and improvements to the original design and gains another 15 years of exclusivity, until the innovation falls into the public domain. This is known in the legal universe as evergreening, a maneuver of prolonging protection over the domain (Paladar 2013). Just in Brazil, 40 records related to the Nespresso system were identified at the National Institute of Industrial Property (INPI) (Paladar 2013).

[12] Founded in 1753, the Douwe Egberts is originally from Netherlands and is dedicated to the processing and marketing of tea and coffee. The company passed to the command of Sara Lee in 1978; and in 2012, with the end of that company, it became was once again an independent company, receiving the name of DE Master Blends 1753. In 2013, the company was acquired by a German investor. A year later, in 2014, the company went through a merger with the coffee business of the American company Mondelez. In Brazil, D.E. Master Blends 1753 owns the brands Café do Ponto, Pilão and Caboclo, among others. For more information: http://www.douwe-egberts.com/, http://www.pilaoprofessional.com.br.

[13] The Ethical Coffee Company was founded by Jean-Paul Gaillard, the executive responsible for the design and consolidation of the Nespresso brand worldwide. He was the director of Nespresso between 1988 and 1997. Soon after, Gaillard left the Swiss company and started to dedicate his time to a new product innovation: to ensure environmental sustainability through capsules with biodegradable material.

[14] Further information can be found at: https://www.dualit.com/products/coffee-capsules-pods.

[15] Two other important competitors are the American Starbucks, Box 5 (Lucas and Simonian 2012), which in addition to the capsules, entered into the machinery business; and Mondelez International, which most recently, in 2014, announced the merger of the coffee division with D.E. Master Blenders 1753. In 2014, Mondelez and D.E. Master Blenders 1753 announced process to unite their coffee divisions (Exame 2014a). The company resulting from the merger, that received the name Jacobs Douwe Egberts, began operating in 2015 (Cafeicultura 2014).

In Brazil, the story was not different, generic brands began invading the market as of 2012 (Isto É Dinheiro 2013). The *Lucca Cafés Especiais*, in 2012, and Café do Ponto, a property of D.E. Master Blenders 1753, which in 2013 launched the brand L'Or, were noteworthy pioneers in this movement. Seleto, Grupo Mex, Três Corações, Utam and Leonardi Cafés, as well as the Portuguese company Delta (Almeida 2013) are important Brazilian companies that participated in this dispute.

Once the competition was established, the price war began, drawn by the new starters looking for their place in the sun.[16] The accessibility of the product to a larger portion of the population perked the market, causing even the emergence of new businesses. For example, companies specialized in reusable capsules - which are filled by the consumer himself with the coffee of their own choice have appeared in the market. The attractiveness of the market has also drawn attention from food retail firms, the case of supermarket chains that managed to launch their own branded capsules. Or even coffee shops, like Starbucks, with investments in brand repositioning (sofa also from home) tailgating the success of the capsules industry. See details in Box 5.

However, the temperature, quantity, speed control and water pressure systems revealed potential applications that transcend coffee. Many investments from this perspective are already real. See, for example, hot soups and broths, relying on Campbell's[17] pioneering spirit, and encapsulated soft drinks, in which Ambev and Coca-Cola[18] are at the forefront of the segment in Brazil.

Only time will tell if the generic capsules have managed to overthrow Nespresso, or whether they will become second-line products

(Isto É Dinheiro 2013).

In fact, in recent years, Nestlé has been losing important legal battles and, although it is still the market leader, its space is getting smaller in the capsule market. According to Euromonitor International, Nespresso's share of worldwide sales in the segment it has created has dropped from 35.8% to 18% in the last 10 years (Peixoto 2016; Le News 2016).

---

[16] The Ethical Coffee Company, for example, was launched on the market with prices reaching three times cheaper than those charged by the Swiss company (Paladar 2013). In Brazil, however, the L'Or brand from Café do Ponto was offered at R$ 1.50 a unit (Isto É Dinheiro 2013).

[17] The capsules will be sold in kits containing noodles and dehydrated vegetables to be mixed into the broth. This contract consisted of a commercial alliance with Green Mountain Coffee Roasters, a company that operates in the capsule coffee segment with its own Keurig brand (Armazém do Café 2013).

[18] AMBEV, in partnership with Whirlpool, manufacturer of Brastemp brand, went ahead, launching in 2015 a pioneer technology in the production of hot and cold drinks. In total, the B.blend machine offers 20 hot and cold beverage options, including tea, juice and soda. The machines are being produced at the Whirlpool factory in Joinville, Santa Catarina state. The capsules are made by Bevys, from Germany, and brought to Brazil. The price of the machine is R$ 3499. The capsules cost from R$ 1.99 to R$ 4.00 (Rodrigues 2015). In parallel, Coca-Cola had already announced, in 2014, the purchase of a 10% stake of Green Mountain Coffee Roasters, the same company that in the US is ahead of the product innovation launched by Campbell's (Uol 2014), see previous note.

The global sales, which grew by 30% per year, decelerated to 10% (US $ 5 billion annually). In Brazil, the second largest consumer of coffee on the planet, Nespresso that held the monopoly now controls 41%. From eight industries producing coffee in capsules in the country in 2014; there are about 100 in 2016 (Peixoto 2016). According to Nathan Herszkowicz, executive director of the Brazilian Coffee Industry Association (Abic), "while in the traditional roasted coffee segment, the profit margin is very low, in the capsule segment it reaches 30%" (Peixoto 2016). As a result, Nespresso business finds itself forced the company to also readjust its prices. Some of its product lines, ranging in 2010 from R$ 1.90 to R$3.00, started to be sold by R $ 1.50 in 2012 (Paladar 2013).

But a close look at how Nespresso reacted to competition reveals that, in the face of this war, the company reinforced its brand strategy based on attributes that seek to offer the consumer a unique experience (Ferreira 2015). Strategies that unfolds in actions coordinated upstream (supply) and downstream (distribution) de product line.

Sooner or later, patents will be lost.
It is up to us (Nespresso) to find the best way to differentiate ourselves, to be unique (Bruegger, Director of Nespresso Américas, available in Mundo S/A (2013).

Nespresso continues to invest in its branded boutiques worldwide (Ferreira 2015), targeting the premium luxury segment, geared towards an accurate niche of opinion makers. Since the 1980s, when the current director at the time, Jean-Paul Gaillard, devised the design of Nespresso, brand differentials were built based on unique consumer experience.

"*The coffee product relates to reason; while the coffee experience provided by Nespresso reaches the emotional*" (Ferreira 2015), based on the development and support of differentiated relationships through its value chain with the consumer, through its boutique concepts.

Gold grains?[19]
🏛 Certified producers, with incentives secured by long-term contracts.
🏛 From all the coffee in the world, 10% is gourmet; and only 1% of it meets the Nespresso standard.
🏛 Coffee worldwide acknowledges its quality (zero defective products, specific sensory profile in the cup, richness and specificity).
🏛 Grain origin control imposed by the company (certified suppliers).
🏛 Cost 20% higher x higher remuneration (on average, 60% higher than those practiced in the market).

---

[19] According to Mundo S/A (2013).

But the unique experience linked to the Nespresso brand also encompasses aspects in terms of unique relationships with its suppliers. Inspired by the gourmetization seen in the wine market, the Nespresso brand goes back to differentials built with suppliers, referring to grain origination strategies, as well as in differentiated relationships in terms of incentive and control, see **Box *Gold grains***.

In this regard, Nespresso develops qualification projects for suppliers in order to meet their differentiated standards. In this direction, there are preferences for long-term contracts with suppliers, who are guided by a code of conduct, to provide practices that respect social and environmental sustainability, while observing good grain quality practices (Saes 2008). The contract also establishes differentiated remuneration for the quality of the grain, which includes the aromatic potential of the beverage, with the valorization of attributes related to the crop and the soil, also going through aspects of harvesting, processing and storage of the coffee beans.

Taking as an example the agreement established between Nespresso and the Cooxupé (Regional Cooperative of Coffee Growers in Guaxupé),[20] located in the south of Minas Gerais state. According to Saes (2008, 2010), the 2006/2007 harvest, 390 members of the cooperative (audited by Imaflora[21]) were part of the program, responsible to provide 1768 thousand 60 kg-coffee bags. From all members, 67.4% are small producers (produce less than 500 bags). In the referred harvest, Nespresso paid a premium price of R$ 15.00 (or 16%) per bag without sifting, which means R$ 40.00 more per bag compared to the average practiced in the local market in the same period (Saes 2008).

Sustained by these differentials and despite the abundant and widespread competition observed in the coffee capsule market throughout the world, Nespresso still represents the largest premium coffee brand in the world (Isto É Dinheiro 2013).

A status that continues to allow Nespresso to enjoy the sweet taste of differentiated price practices with respect to the competition. In fact, the products that suffered price adjustment essentially refer to Nespresso's front line, maintaining price differentials to their premium products (selections and special editions, for example), in relation to other companies (Isto É Dinheiro 2013).

In addition, Nestlé also seems to use Dolce Gusto as a protective shield for Nespresso. Although not conceived as a second line, preliminary indications suggest this role to Dolce Gusto, who should follow the price position of the main competitors in the Brazilian retail, among them are Três Corações and Utam.

Therefore, there is an incentive for future studies, particularly directed to the investigative analysis of the price evolution practiced by the company, in its two brands of capsules, along with the movement of new companies into the market.

In any case, it is believed that the Nespresso case provides elements to the argument for greater robustness of the outside perspective as a basis of a less ephemeral competitive advantage, when compared to differentiation strategies based exclusively on inside actions.

---

[20] Complementary information about the cooperative can be found on their website: https://www.cooxupe.com.br/.

[21] The Institute of Forest and Agricultural Management and Certification. Available at: http://imaflora.blogspot.com.br/.

Through this paradigm, the supplier ceases to be understood as interchangeable (in which cost reduction prevailed), assuming a crucial role in creating and sustaining competitive advantages for the company, given that a considerable part of the value or quality of the final product is derived from its origin and processing. This context gives the industry the opportunity to differentiate products by means of social and environmental attributes.

References

Agência Estado. (2013). *Mondelez vai lançar cápsulas de café para máquinas da Nespresso.* Retrieved from http://g1.globo.com/economia/agronegocios/noticia/2013/06/mondelez-vai-lancar-capsulas-de-cafe-para-maquinas-da-nespresso.html.
Almeida, M. (2013). *Cresce disputa de fabricantes de máquina de cafés pelo mercado de cápsulas. Ig São Paulo.* Retrieved from https://economia.ig.com.br/empresas/2013-08-20/cresce-disputa-de-fabricantes-de-maquina-de-cafes-pelo-mercado-de-capsulas.html.
Armazém do Café. (2013). O gigante vai acordar? Retrieved from.
Barney, J. B. (1991). Firm resource and sustained competitive advantage. *Journal of Management, 17*(1), 99.
Blecher. (2014). *Nestlé vai abrir fábrica de cápsulas de café em Minas. Globo Rural.* Retrieved from https://revistagloborural.globo.com/Noticias/Agricultura/Cafe/noticia/2014/12/nestle-vai-abrir-fabrica-de-capsulas-de-cafe-em-minas.html.
Café Point. (2013). *Europa revoga patente de máquina Nespresso.* Retrieved from http://www.cafepoint.com.br/noticias/mercado/europa-revoga-patente-de-maquina-nespresso-86540n.aspx.
Cafeicultura. (2014). *Mondelez investe US$ 50 milhões em fábrica de cápsulas de café.* Retrieved from https://revistacafeicultura.com.br/index.php?tipo=ler&mat=55725.
Exame. (2014a). *Nestlé lança em Minas pedra fábrica de cápsulas de café.* Retrieved from https://exame.abril.com.br/geral/nestle-lanca-em-minas-gerais-pedra-fundamental-de-fabrica-de-capsulas-de-cafe/.
Exame. (2014b). *Mondelez e D.E. Master Blenders formam gigante do café.* Retrieved from http://exame.abril.com.br/negocios/noticias/mondelez-e-d-e-master-blenders-formam-gigante-global-do-cafe.
Ferreira, R. (2015). *Como a Nespresso se tornou um ícone sem nunca ter vendido café. Endeavor.* Retrieved from https://endeavor.org.br/marketing/valores-intangiveis/.
Food Service News. (2014). *Grandes lucros com a Páscoa.* Retrieved from https://www.foodservicenews.com.br/grande-expectativa-para-pascoa/.
Henson, S., Humphrey, J. (2009). The impacts of private food safety standards on the food chain and on public standard-setting processes, Joint FAO/WHO Food Standards Programme, Codex Alimentarius Commission, Third-second Session, Rome.
Humphrey, J., Memedovic, O. (2006). *Global value chains in the agrifood sector.* Retrieved from http://tinyurl.com/y8qh4obd.
Isto É Dinheiro. (2013). *Chegou o genérico do Nespresso.* Retrieved from https://www.istoedinheiro.com.br/noticias/negocios/20130130/chegou-generico-nespresso/386.shtml.
Le News. (2016). *Nestle's coffee business is competing with itself.* Retrieved from https://lenews.ch/2016/06/30/nestles-coffee-business-is-competing-with-itself/.
Lopez, J. (2014). *Changing role of business. Creating Shared Value Forum 2014.* Retrieved from https://www.youtube.com/watch?v=SOz5Qw1kc84.

Lucas, L., Simonian, H. (2012). *Nestlé loses bid to stop rival coffee capsules. Financial Times.* Retrieved from https://www.ft.com/content/062146de-e793-11e1-8686-00144feab49a#ixzz3g 3qSm4Tb.

Mintel. (2016). *Global food and drink trends.* Retrieved from https://www.mintel.com/presscentre/ food-and-drink/mintel-identifies-global-foodand-drink-trends-for-2016.

Mundo S/A. (2013). *Fábricas apostam nas cápsulas de café espresso para vencer disputa no mercado consumidor. Globo News.* Retrieved from https://www.youtube.com/watch?v= zTjx91Ck0MM.

Nestlé. (2008). *Our company.* Retrieved from http://www.nestle-nespresso.com/about-us/ our-company.

Paladar. (2013). *Patente encapsulada. Paladar (Estadão).* Retrieved from https://paladar.estadao. com.br/noticias/bebida,patente-encapsulada,10000010207.

Peixoto, F. (2016). *O gosto amargo da Nespresso. Isto é Dinheiro.* Retrieved from https://www. istoedinheiro.com.br/noticias/negocios/20160303/gosto-amargo-nespresso/347086.

Portugal, M. (2013). *Kopenhagen lança ovo de Páscoa com cristais Swarovski. Exame.* Retrieved from https://exame.abril.com.br/marketing/kopenhagen-lanca-ovo-de-pascoa-com-cristais-swarovski/.

Rodrigues. (2015). *AMBEV e Whirlpool criam empresa para lançar máquina de bebidas em cápsula. O Globo.* Retrieved from https://oglobo.globo.com/economia/ambev-whirlpool-criam-empresa-para-lancar-maquina-de-bebidas-em-capsula-16202443#ixzz3g49wJtyP.

Rogê, L. (2014). *Chocolate: Um mercado disputado. Exame.* Retrieved from https://exame.abril. com.br/blog/investidor-em-acao/chocolate-um-mercado-disputado/.

Rosa. (2013). *E agora, para onde vai o café? Paladar (Estadão).* Retrieved from https://paladar. estadao.com.br/noticias/bebida,e-agora%2D%2Dpara-onde-vai-o-cafe,10000010199.

Sá, S. (2012). *Nestlé lança Suflair Dark em edição limitada. Exame.* Retrieved from http://exame. abril.com.br/marketing/noticias/nestle-lanca-suflair-dark-em-edicao-limitada.

Saes, M. S. M. (2008). *Estratégias de diferenciação e apropriação da quase-renda na agricultura: A produção de pequena escala.* Tese de Livre Docência: Universidade de São Paulo, Departamento de Administração FEA.

Saes, M. S. M. (2010). Rent appropriation among rural entrepreneurs: Three experiences in coffee production in Brazil. *RAUSP: Revista de Administração da USP, 45*(4), 313–327.

Saes, M. S. M., & Silveira, R. L. F. (2014). Novas formas de organização nas cadeias agropecuárias brasileiras: Tendências recentes. Estudos Sociedade e Agricultura. *Rio de Janeiro, 22*(2), 386–407.

Silva, V. L., Kasemodel, M. G. C., Makishi, F., Souza, R. C., & Santos, V. M. (2019). Beauty is not only skin deep: Food ethics and consequences to the value chain. *The Interne, 14*, 286–303. https://doi.org/10.18568/internext.v14i3.489.

The Brand Gym. (2012). *Espresso Wars–Starbucks takes on Nespresso.* Retrieved from https:// thebrandgym.com/espresso-wars-starbucks-takes-on-nespresso/.

Uol. (2014). *Coca-Cola fecha negócio e deve lançar cápsulas para fazer bebida em casa. Uol Economia.* Retrieved from https://economia.uol.com.br/noticias/redacao/2014/02/06/ coca-cola-vai-lancar-capsulas-para-fazer-bebida-em-casa.htm.

Chapter 5
Rules of the Game

"When the consumer selects a product, she chooses an entire value chain, compacting with the governance of relationships."

Foss (2013)

Nicolai Foss, Professor at Copenhagen Business School.

Abstract If a classic question of the organizational world refers to deciding to differentiate (or not), a derivation of this aspect is that social drivers linked to food technology lead companies to account for quality issues that transcend the processing limits, incorporating well-being and empowerment of producers. And with that the need to review the bases of selection and development of suppliers, as well as the mechanisms of incentive and control. Discussing these managerial implications is the goal of this Chapter. Although the classical and well-known discussion in the applied social sciences, as in business and economics, this reflection remains almost marginalized in the mainstream of food science and technology.

Keywords Organizational economics · Business economics · Food studies · Relationship governance · Food value chain · Food value adding · Consumption drivers · Brand value · Brand information · Sustainable competitive advantage · Human resources training

5.1 A Matter of Relationship

Differentiate or not?! It is a classic question from business economics (Porter 2005; Besanko 2006). Nowadays with an updated context, as previously discussed in the previous Chapters. The ongoing differentiation movement we are currently experiencing transcends the boundaries of the industry, with consequences through the value chain in which food processing is located as well as in the relationship governances, in order to ensure that the new and more complex information elements associated with the brand are met.

© Springer Nature Switzerland AG 2020
V.-L. Silva, *Social Drivers In Food Technology*,
https://doi.org/10.1007/978-3-030-50374-1_5

The organization of the food systems, in terms of their coordination and governance, undergoes a profound restructuring in fact,[1] to which Azevedo and Silva (2014) attribute a series of elements, such as technological breakdowns, different patterns of competition and consumption trends. This context has marked the dynamics of the food processing sector we have discussed in Chap. 1.

From its conception, in the early nineteenth century, to the present day, the distinctive information signal transmitted by its brands has undergone extraordinary transformation, proving to be significantly more complex in consonance with the evolution through which social role of processed food has passed, mainly in terms of the new drivers associated with their consumption, Fig. 5.1.

The relationship between these new consumption drivers and the reorganization of food systems is related to the greater complexity of the information transmitted by the brand. Once, The food industry is more and more stimulated[2] to differentiate its brands in a dual-level: physical health of the consumer; at the same time of the environment and society; what goes through differentiating elements built outside, referring to throughout the food system. Figure 5.2 summarizes some of the main trends that follow this path.[3]

Fig. 5.1 Consumption drivers and complexity of the information transmitted by the brand of processed food

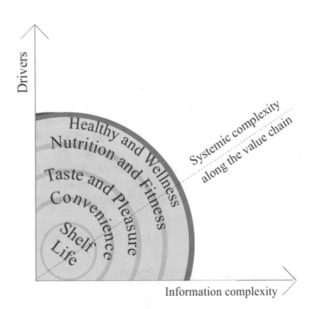

---

[1] The way food differentiation strategies can affect the value chain governance structure was originally discussed by Ménard (1996) for the case of high-quality broiler chickens, and by Farina and Zylbersztajn (1997), for the case of gourmet coffees (Azevedo and Silva 2014).

[2] Process that finds in the consumer one of its main drivers but is not restricted to it. In this process of new demands at the table, the role of other agents cannot be marginalized, such as the Government itself and its regulations.

[3] Obviously, this argument cannot be generalized when dealing with the different growth vectors of the contemporary food industry. However, this represents an equally important part of the market, referring to consumers (with better access to information and greater purchasing power) who have already met basic conditions (food safety), as well as satisfied demands related to taste, practicality and convenience, they now demand for health and wellness on the table.

INDIVIDUAL	ENVIRONMENT	WELLNESS
Physical e Mental	*Environment*	*Society*
- Fortified (e.g. vitamins, minerals, collagen/omega 3) - Sugar-free - Naturally sweetened - Based on selected natural ingredients - Artificial aroma - Fat-free/low-fat content - Sodium free/low sodium content - Cholesterol free/enriched with ingredients that act against cholesterol - Free from allergenic ingredients (e.g. gluten, lactose) - Free from pesticides	- GMO-free - Green certificates (e.g. ecologically adequate management, biomaterials 100% recyclable) - Lower CO_2 emission rate - Environment friendly packaging (size reduction, use of new materials, 100% recyclable, new printing processes (vegetable-based ink) lower pollutant rates) - Environmental certificates, such as Rainforest and RSPO (free palm oil) - Responsible industry (donagem for specific programs supported by the company)	- Origin of raw material - Social certifications, such as family agriculture - Sustainable agriculture and fishing - Fairtrade - Animal welfare - Local development programs (community growth)

Fig. 5.2 Health differentiation movement registered in food marked (Note: *GMO* genetically modified organism, *RSPO* roundtable on sustainable palm oil)

A question that remains open is if this movement is predominantly an important gateway to new and small entrants, which in turn forces the brand repositioning of historically consecrated firms. This is a discussion that opens a research agenda, starting with the investigation of the organizational determinants to investing in outside actions for brand differentiation; and the relevance of small businesses on this choice.

But what are the management implications of the decision between limiting to an inside approach or harmonize with differentiating attributes built outside the technological boundaries? This will be discussed next.

5.2 From Farm to Fork

The organizational capacity to grow and survive in the market, whether recurrent or new (Farina 1999), necessarily involves systemic relations from farm to fork in the value chain in which the processing company operates. Take or instance the threats to business strategies due to vertical coordination bottlenecks, such as inefficiency of suppliers to achieve requirements (cost, time, quantity, quality), or problems associated with distribution and marketing (Azevedo and Silva 2014).

The orientation of this articulated action tends to be based on optimizing production and operations, as well as on minimizing of both production and transaction costs. This is because the differential ensured within the boundaries of the company allows product differentiation to be based on quality standards, generally disseminated in the industry sector, with measurable and easily controlled attributes. A basis that favors the adoption of tenuous mechanisms of incentive and control, such as the *spot* market, or even short and long term contracts, which shows interdependence between low complexity agents, such as joint or sequential, see Table 5.1.

In summary, as discussed in Saes (2008) and Saes and Silveira (2014), a joint interdependence means an arrangement characterized by the acquisition/sale of commodities. Horizontal and vertical relationships are sparse and social ties are weak. Prices reflect the full range of incentives required. An example of this type of interdependence is the characteristic relationship among milk producers and dairy product industry in the animal protein commodity market.

On the other hand, sequential interdependence refers to the organization of rural producers in response to the specific strategy of the processing company, to guarantee the product supply (volume and price) or the standards associated with the brand (quality attributes) (Saes 2008; Saes and Silveira 2014).[4] See, for example, a typical

Table 5.1 The value-adding process inside x outside: systemic game of distinct complexities

Orientation	Emphasis	Competitive advantage bases (CA)	Resource profile	Types of independence	Dominant governing	Governing instrument	Supply chain impact
INSIDE Value-adding assured within company boundaries (product and process engineering)	Homogeneity (final product standardization)	-Product -Process ↓ -Ephemeral CA (easily reproduced by competition)[1]	-Acquisition of commodities or products with specific quality standards. -Attributes disseminated through the market or specific to a brand strategy from the processing company, but easy to measure or control.	-Joint or sequential	Weak incentive and control mechanisms: - Spot market - Short or long term contracts	-Market price (spot) -Price deermined by the processing company (equal or greater them market prices) -Different types of long term contracts (relational or formal) for supply (volume) under the processing company's authority	-Lower complexity systemic game -Weak social ties -Transactional] -Perspective (emphasis: minimizing transaction costs; optimizing production and operations) -Suppliers understood as "exchangeable"[]
OUTSIDE Value-added built along the chain (relationship engineering)	Authenticity (genuine raw material and final product)	-Relationship ↓ -Sustainable CA (competitive differentials hold in time)	- Acquisition of product with a specific quality standard that also presupposes specific vertical and horizontal actions - Attributes that are difficult to measure and control (goods of belief)[3]	-Reciprocal Allied to Sequential	More incentive and control mechanisms: -Long term contracts (relational or formal)	-Price set by the company -Authority (exercised by certifying institution) -Long term (formal or relational) contracts with the processing company, backed by trust in the relationship -Knowledge expertise (mutual adjustments)	-More complex systemic game - Strong social ties -Tends to enable better income appropriation by rural connections -Sustaining relationships with suppliers are understood as strategic to maintain differentials

Notes: [1]Unless protected by patent. [2]Stevenson and Pirog (2013). [3]Barzel (1982). Source: Based on Saes and Silveira (2014), and Lazzarini et al. (2001).

---

[4] According to Saes and Silveira (2014), sequential interdependence denotes an arrangement in which tasks are structured sequentially, that is, one activity or one agent of the company precedes the other, generating co-specialization. Also, Saes and Silveira (2014) considers co-specialized assets when these prove to be comparatively more valuable in combination than in isolation; and bring competitive advantage to the company they are part of. When specialization results from a relationship between firms, the one that holds the residual property rights is the one that appropriates most of the value generated (Saes and Silveira 2014).

relationship established between the milling industry and cocoa producers. As seen in other sectors, such as orange and sugar cane, the sequential arrangement is formed so that the cocoa processing company can guarantee the supply of its products.

Another example is the arrangement established between the baby food processing leader in Brazil and its carrot suppliers. The relationship requires the cultivation of a specific variety. In addition, to guarantee the necessary volume of supplies, the arrangement is designed to ensure the minimization of production costs as a result of the higher industrial yield originating from the variety of the input in question.

Sequential interdependence is also expected when seeking to guarantee specific attributes (objective, easy to measure and to control) associated with the strategy of the brand. An example of sequential interdependence in the context is the roasted snacks, which requires a specific variety of corn in sensory terms.

In spite of the main motivation, in the sequential arrangement, the hierarchy-based authority is necessary to avoid the strategic information (as bases of differentiation associated with the brand) to be dispersed (Saes 2008; Saes and Silveira 2014). This can be translated into the use of long-term (formal or informal) supply contracts (with the specification of volume and price). Just as in joint interdependence, under the sequential arrangement, social bonds are also weak; and although using stricter forms of governance (procurement contracts), suppliers continue to be understood as interchangeable (Pirog 2013) under the most search for minimizing transaction costs and optimizing production and operations.

The problem gets even more widened when there is reciprocal interdependence[5] allied to the sequential-basis of the outside differentiation perspective- in which there is the acquisition of inputs with specific quality but presupposes collective actions among producers (Saes 2008; Saes and Silveira 2014). This is the case of organic processed food, based on the indication of origin or also certification as promoters of impact relationships (socio-environmental).[6] Drawing a parallel line with the previous examples, the positioning of the baby food or snacks brand is once again suggestive, based on a specific variety of input, seeking to address sensorial issues, with the differential of organic certification for the rural suppliers (and/or other seals, such as fair trade programs, local development, family farming or animal welfare).

Intangible attributes are usually related to a rare, valuable, non-replaceable and hardly copied resource. And so, its use as a basis for brand differentiation guarantees a differential that is difficult to copy by the competition (golden dream of competitive war); namely a sustainable competitive advantage (i.e., kept in time) (Barney and Firm 1991). Which in turn, however, inevitably culminates in a systemic game

---

[5] Arrangement involving (horizontal) relations between the parties, and the input of one agent depends on the input of the other and vice versa (Saes and Silveira 2014).

[6] Such an arrangement affects the subsequent relationship as it allies to sequential interdependence with the downstream and upstream segments of the value chain. Since it deals with a relationship that produces synergies, reciprocal interdependence also enables co-specialization (Saes and Silveira 2014).

of higher complexity: greater impact on the relationship governance and inevitable reorganization of the productive food systems.

Under the reciprocal interdependence allied to sequential interdependence, each producer is mutually dependent on the choices and actions of the others. The rights of decision-making are distributed among rural producers, implying in a complex solution process, with the risk of free-rider behavior[7] (Saes 2008; Saes and Silveira 2014).

Since this problem is not trivial, some mechanisms of consensus, negotiation among the parties and mutual adjustment are assumed. The coordination and adaptation forms require learning through feedback and trust rather than central decision-making planning (Saes and Silveira 2010).

Horizontal relational contracts (trust) are expected between producers; and long-term (relational or formal) contracts with processor industries mediated by third parties (certification) (Saes and Silveira 2010). In addition, the maintenance of the relationships between processing companies and rural producers is understood as strategies to keep the differentials linked to the brand, building up trust (Saes and Silveira 2010).[8] A conceptual model about the perspectives of value-addition (inside versus outside) and the consequences in the coordination of the value chains in which the processing companies are inserted are summarized in Table 5.1.

Closing this chapter, the purpose of the next two sections is to bring empirical elements to the discussion, using the partnership established by one of the largest food processors in Japan, Meiji, in the supply of cocoa from the Amazon Forest, by a cooperative of Japanese immigrants, CAMTA (Portuguese acronym for *Cooperativa Mista de Tomé-Açu*). A partnership that runs away from the dominant governance in the chocolate value chain; and which suggests the power[9] to build brand differentiators outside the box.

5.3 Recalculating the Route

The chocolate value chain can be characterized into four levels. The first one, inside the farm, refers to the handling, production, and processing of the almonds. The liquidity of cocoa in the market (the reality of informal commerce is verified by the street commerce), coupled with the extractive tradition rooted in the sector (family income depending on a few native cocoa trees) provides means to turn the crop a reality in small and medium properties.

---

[7] When an economic agent benefits from a certain advantage without incurring the costs of the related good.

[8] This discussion suggests a new perspective of analysis on value distribution issues. Suggestive to consider industrial optics, complementary to the usual activities of the rural producer.

[9] As pronounce at the opening of this chapter is not (only) about organizational separation. This strategy can actually represent attractive business to organizations.

In Brazil, 91% of the properties dedicated to the activity are small (less than 100 ha), occupying only 1% of the area destined to the activity in Bahia. However, the medium-sized properties (from 100 to 500 ha) represent 8% of the total number of rural properties in Brazil, accounting for 72% of the total area (Cuenca and Nazário 2004).

After going through the fermentation, drying and pre-classification processes in the rural properties, the almonds are commercialized to the milling industries. The dissemination of small intermediaries is observed, buying the almonds directly from the farmers and passing them on to the industries. Frequently these intermediaries settle in smaller municipalities to receive almonds from small producers.

In the second level of the value chain (milling industry), the first industrial transformation of the almond is performed, extracting the mass of cocoa, a key ingredient in the chocolate production. Two by-products are obtained by pressing the mass of solubilized cocoa: cocoa butter (liquid part, being the noblest by-product) and cocoa cake (solid portion).

The third level of the value chain is the chocolate topping industry (dedicated to the foodservice sector), where the cocoa mass is processed by adding cocoa butter or another type of fat, such as hydrogenated, in addition to other ingredients, such as sugar, milk or soy lecithin, according to the formula and standard of the manufacturer.

Lastly, the fourth level is responsible for the modeling of chocolate in the format of tablets, candies and confectionery.

Traditionally, the chocolate value chain is controlled by the milling and processing industries using the spot market or supply contracts in order to ensure the volume and price of the raw material.

In the classic coordination model of the chocolate value chain, the greater bargaining power of the milling industry is favored by the small scale and geographic sprawl of rural properties, demonstrating tenuous incentives for better product quality (by offering bonuses on top of the market price).[10]

In this respect, three factors deserve attention. This first of them is the absence of criteria (transparent to the producer) that allows to assess and fit the organoleptic quality characteristics of the almonds. Second, the strategic positioning characteristic of the sector focused on price. Third, relationships between almonds sellers and mills are essentially informal but reproduced over time.

In this context, reciprocal fidelity contributes to minimizing the coordination problem, given the corner where the sector is inserted (discrepancy between cocoa supply and demand for its derivatives). However, market transactions are also

---

[10]Aggravating the problem, the shortage of raw material in the market and the liquidity characteristic of cocoa, even for low quality, contribute to the commercialization of healthy almonds mixed with those damaged by the fungus, sometimes without going through the adequate processes (fermentation and drying). In the conventional chocolate market, actions inside (processing) make it possible to mask variations (even drastic) in the quality of the raw material. In fact, the quality of the final product, in this case, seems to depend little on attributes associated with the raw material.

influenced by short-term uncertainty due to the unpredictability of the supply[11] (Saes et al. 2014).

Changes in this classic coordination pattern begin to be seen from the new consuming boosters that influence the appetite of the contemporary consumer, who goes through a more complex characteristic attributed to the social role of the processed food.

In this case, the chocolate market opens up opportunities for differentiated products in terms of disseminating new bases of relationship with the rural link, unfolding in new forms of value chain governance.

Relationships that also tend to favor differentiated sensorial attributes, potentializing the bases of product differentiation. At the extremity of this discussion is the case of the relationship between Meiji and CAMTA.

5.4 'CAMTA'ste Chocolate?

Fine cocoa means raw material (almond) obtained respecting specific handling, harvesting, post-harvesting and processing practices (fermentation and shelling). Process differentiation that restricts obtaining fine chocolate.

However, the basis of this product differentiation can be exponentially expanded when associated with other tacit elements: for example, the cocoa-nut extracted from the Amazon Forest,[12] based on the agroforestry system,[13] and the activities of Japanese immigrants, gathered in a cooperative, which have conquered these lands and settled there since the beginning of the twentieth century. This, for example, was the treasure found by one of the largest Japanese food processing company firms, Meiji Seika[14] at CAMTA (*Cooperativa Agrícola Mista de Tomé Açu*, in Portuguese).

Prior to its partnership with CAMTA, Meiji had already partnered with cocoa producers in 20 countries, including Ghana, in Africa, where the company used to acquire around 20,000 tons of almond/year. However, that search ended in 2009, when Meiki Seika formalized an exclusive commercial partnership with the cooperative (Saes et al. 2014).

Environmental sustainability is a necessary but not sufficient condition to produce differentiated cocoa, in sensory aspects. Indeed, the early years of CAMTA'

---

[11] The biannual aspect, characteristic of perennial plants, such as cocoa, introduces some unpredictability in the supply. There are reasonable doubts about the capacity of the producers in traditional regions to invest in the renewal of cocoa that are reaching the end of their productive life.

[12] As discussed by Saes et al. (2014), the cocoa production in Brazil is concentrated in the southern region of Bahia. But the Amazon states, especially Pará, are challenging this tradition, revealing a new geography of cocoa production in the country. As one of the protagonists of this story, there is the cooperative CAMTA, which in 2013 produced 600 tons of cocoa, approximately 0.5% of Brazilian production (Saes et al. 2014).

[13] Sustainable agriculture.

[14] In 2012, Meiji Seika went through a merger with another Japanese giant, Nyugyo. This merger has helped the company position itself as the largest food processor in Japan.

relationship with Meiji were marked by problems, due to the quality of the almond supply (especially in terms of the fermentation process). Fact materialized already in the first delivery, not attending the required standards. The output was the long-term investment of both parties, Meiji and CAMTA, in the generation and dissemination of knowledge. Meiji sent to Brazil researchers who started a campaign to develop good fermentation practices with the producers. Two years (2009/2011) of investments in research and extension were necessary to reach the desired cocoa quality, in terms of almond fermentation (Saes et al. 2014).

CAMTA, however, provided the area for cocoa production and the infrastructure to conduct fermentation tests. In addition, the cooperative, through its directors, played a decisive role in the involvement and engagement of producers in the new practices. One of the directors of CAMTA, then municipal secretary of agriculture at Tomé-Açu and producer of cocoa in the region, provided raw materials, infrastructure and labor for research (Saes et al. 2014).

The first fruits were harvested in 2010, when the variety of cocoa produced by the cooperative (type C-27) was awarded at the International Cocoa Award Excellence in France, placing the cocoa produced by CAMTA among the best cocoas in the world. On that occasion, Meiji launched a specific line of chocolates using CAMTA's Agroforestry System for marketing appeal (Saes et al. 2014).

The great value differential attached to CAMTA's product is based on this context. Launched in 2011, 'Agroforestry Chocolate' carries on its packaging the origin specification, 100% cocoa from CAMTA,[15] with the declared identification of the producers whose almonds were processed.

Contractually, the agreement stipulates 150 tons/year over 10 years. The first batch delivered in 2009 presented 100 tones. In 2010, the stipulated amount of 150 tons was reached, from which 24% of CAMTA's production. The expectations for the following years were to reach 300 tons/year. CAMTA buys cocoa from all members of the cooperative and performs the classification during the delivery by producers. The product whose quality does not reach the standard is sold in the domestic market, being its main buyer the miller Delfi Cacau Brasil Ltda., responsible for absorbing around 60% of CAMTA's production (Saes et al. 2014).[16]

---

[15] Founded by Japanese immigrants in the midst of the Amazon rainforest, in 1931. The origins of the Cooperative Agrícola Mista de Tomé Açu (CAMTA) refer to Japanese immigration to Brazil in the early twentieth century. The first records of Japanese immigration in Brazil dates back to 1908, with the arrival of the first ship bringing 780 immigrants. In 1920, there were already around 24 thousand immigrants. In the 1930s, another 80,000 Japanese people crossed the oceans towards Brazil (Saes et al. 2014). The first immigrants arrived in the region (Vale do Acará, Northeast of Pará state, about 200 km from the capital, Belém) two years earlier, in 1929. The immigration to the region occurred through a Japanese-Brazilian agreement that seemed to be a solution for both countries: Brazil needed labor to exploit the Amazon region and Japan had excess farmers living in precarious conditions (Saes et al. 2014).

[16] Initially, the cooperative should deliver all production to CAMTA. But the Cooperative makes this restriction flexible since there was a need for some producers to receive the production in advance. There are many buyers of cocoa in the region. In the cooperative, the member is payed after the product is commercialized. In the case of selling to Meiji, payment is made before the product arrives in Japan, around 1–2 weeks after the sale (Saes et al. 2014).

References

Azevedo, P., & Silva, V. L. S. (2014). *Indo além da relação franqueador-franqueado. In: Teoria e prática do franchising: Estratégia e organização de redes de franquias.* São Paulo: Atlas.

Barney, J., & Firm, B. (1991). Resource and sustained competitive advantage. *Journal of Management, 17,* 1.

Barzel, Y. (1982). Measurement cost and the organization of markets. *Journal of Law and Economics, 25.*

Cuenca, M. A. G., & Nazário, C. C. (2004). *Importância econômica e evolução da cultura do cacau no Brasil e na região dos Tabuleiros Costeiros da Bahia entre 1990 e 2002* (p. 25p). Aracaju: Embrapa TC. Retrieved from http://www.cpatc.embrapa.br/publicacoes_2004/doc-72.pdf.

Farina, E. M. M. Q. (1999). Competitividade e coordenação de sistemas agroindustriais: um ensaio conceitual. *Gestão & Produção, 6*(3), 147–161.

Farina, E. M. M. Q., & Zylbersztajn, D. (1997). *Deregulation, chain differentiation, and the role of government. 1st Brazilian workshop on agri chain management.* Ribeirao Preto: FEA/USP.

Foss, N. (2013). *Collective motivations and the theory of the firm.* Speech during the VIII Research Workshop on Institutions and Organizations Conference (RWIO). 013. Personal records.

Lazzarini, S. G., Chaddad, F. R., & Cook, M. L. (2001). Integrating supply chain and network analysis: the study of netchains. *Journal of Chain and Network Science, 1*(1), 13–22.

Ménard, C. (1996). On clusters, hybrids, and other strange forms: The case of French poultry industry. *Journal of Institutional and Theoretical Economics, 152,* 154–195.

Saes, M. S. M. (2008). *Estratégias de diferenciação e apropriação da quase-renda na agricultura: a produção de pequena escala.* Tese de Livre Docência: Universidade de São Paulo, Departamento de Administração FEA.

Saes, M. S. M., Silva, V. L., Nunes, R., & Gomes, T. M. (2014). Partnerships, learning, and adaptation: A cooperative founded by Japanese immigrants in the Amazon rainforest. *International Journal of Business and Social Science, 5,* 131–141.

Saes, M. S. M., & Silveira, R. L. F. (2014). Novas formas de organização nas cadeias agropecuárias brasileiras: tendências recentes. Estudos Sociedade e Agricultura. *Rio de Janeiro, 22*(2), 386–407.

Stevenson, G.W., Pirog, R. (2013). *Values-based food supply chains: Strategies for agri-food enterprises-of-the-middle. Working paper.* Retrieved from http://www.cias.wisc.edu/wp-content/uploads/2013/04/valuechainstrategiesfinal072513.pdf.

Chapter 6
Food for Thought

Abstract From one extreme to the other of the evolution that has been witnessed throughout the two centuries of the contemporary food industry history, the quest for safe and healthy food continues to pull industry development and scientific knowledge, but under a new complex context. This chapter is precisely concerned with what can be expected to follow in this instigating story, instigating considerations to the importance of looking out-of-the-box, which is believed to be of great relevance to the continued process of knowledge generation and human resource formation.

Keywords Teaching · Food · Social drivers · Food ethics · Consumer behavior · Food studies · Information costs theory · Quality standards · Relationship governance · Food chain value · Human resources training

6.1 Visible Hidden

This book ends by recalling the emblematic phrase by Prof. Paul Singh, in 2012, addressed to young university students in the area of food science and technology, during a lecture given at the University of Porto:[1]

> When it comes to food engineering you have to look outside the box.
> There are lots of opportunities right at the boundary (Singh 2012).

The out-of-box context challenges the food industry while challenging history-driven businesses. The future is yet to come. But the present already reveals new chapters in the history of the food industry development, with important management unfolding regarding knowledge generation and the formation of human resources.

Undeniably, *"the merit of the food industry of the twentieth century was scaling-up processes developed by artisans into fabrication lines that consistently produced thousands of units per hour of microbiologically safe, nutritious and appealing foods"* (Aguilera 2006: 1147).

---

[1] See at: https://www.youtube.com/watch?v=23eqfw2aaI8.

Nevertheless, making an analogy here with what was discussed in Chap. 1, the war is far greater since consumer remains dissatisfied.

Indeed, *"food processing is facing a major challenge in the sense that the food chain is reversing and now it is the consumers who tell producers what they want to eat. This global tendency is reshaping the industry into one that provides, in addition to safe and high-quality foods, products that contribute to the health and wellness of consumers"* (Aguilera 2006: 1147). In practical terms, this trend means that several existing products have to be redesigned and new products will need to be created to satisfy this demand (Aguilera 2006: 1147).

Kipping this in mind, in the literature we find a growing discussion about where we are heading in terms of the future of scientific knowledge in food processing and engineering. A hot topic encouraged by the seminal study co-signed by I. Sam Saguy, R. Paul Singh, Tim Johnson, Peter J. Fryer and Sudhir K. Sastry.

In a delightfully provocative reading, Saguy et al. (2013) invite us to think about where we are, in the best or worst times of scientific knowledge in food processing and engineering. Which brings us another provoking reflection: ***how far are we from consumers?***

That's because we have to overcome the Narcissus complex and not be short-sighted by the fascination of technology. As Saguy et al. (2013) stated: ***"If a process is fascinating and attractive to academics it does not make it capable of producing products that improve their marketability or value through the consumers' eyes"***. See **Box *Far from*.**

Trying to not let the glare of the fascination for technology blind our eyes, it is fundamental to remark *"people buy food for the benefits they provide and how they fit into personal and societal values, such as sustainability and environmental concerns"* (Weaver et al. 2014: 1527). In a conception that encompasses adding food product value derived from social and environmental issues of the relationships established along the entire value chain.

As highlighted by Lopez (2014): *"Food product has a lot of value (quality), also referring to where (and how) it comes from"*. Reprogramming the industry, following a business sense. There is no escape from this new route in which food ethics represents 'the new norm'. take a look in Chaps. 2, 3 and 4.

However, it turns out that this reprogramming poses an extra problem for the sector. A true story behind results in an unprecedented information problem for the food industry (Mintel 2016).

It is no longer about commodities or experience goods. In the sense of products that before or during consumption could have their quality standards obtained by laboratory analysis or even sensory experiments. The quality standards related to food ethics are (also) based on attributes understood as credence goods. In which even after consumption, proving such patterns becomes impossible. Child slave labor, and deforestation. For example, something that haunts (still) processors. Or scandals related to the commercialization of conventional foods as if organic or fair trade. It is a consecrated economic problem associated with the nature of a belief asset and the Information Costs Theory impacting the organization of the markets (Barzel 1982).

Far from

How far are we from consumers? Inspired by that golden question, Kasemodel et al. (2016) wonder if it is possible to identify a trail of crumbs concerning consumer behavior in the Food Science and Technology field. And, if that trail exists, they ask to which direction it is leading academia to in terms of research trends.

By following a systematic review strategy, a total of 1786 articles from 1993 to 2013 were analyzed. A recent increasing concern regarding consumer behavior was evident. Nevertheless, it is quite far from the main interest of researchers. Despite an apparent evolution, only 1.11% of all articles published in 2013 directly cited terms related to consumer behavior. With the USA and Spain having a significant role in leading the way. Eight other countries that exhibited similar influences are: Italy, England, Australia, Germany, Denmark, France, Netherlands and Brazil (Kasemodel et al. 2016).

Evolution in the number of articles referring to consumer behaviour.

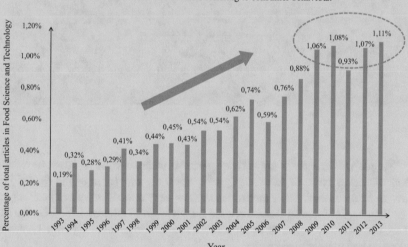

Evolution in the number of articles referring to consumer behaviour. Source: Kasemodel et al. (2016).

Among the most popular topics, Kasemodel et al. (2016) have observed seven major keywords: sensory, health, safety, willingness to pay, packaging, ethics, and lifestyle/convenience. While it is true that the results suggest that publishing trends (hot topics) were influenced by where the research was conducted; suggesting that scientific knowledge does not occur in a vacuum.

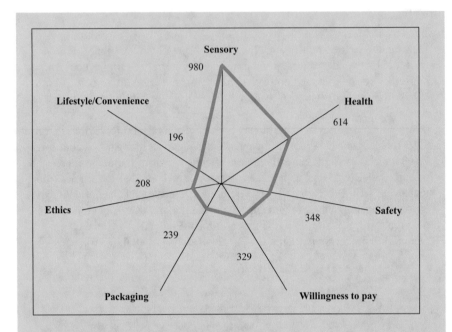

Hot topics in reference to a general global pattern. (Source: Kasemodel et al. (2016))

"People will exchange only if they perceive what they get to be more valuable than what they give. To form such perceptions, the attributes of the traded items have to be measured. Some measurements are easy to obtain; others pose difficulties. Yet what is observed is seldom what is truly valued" (Barzel 1982). On the basis of the economic problem, three aspects are seen, informational asymmetry, adverse selection and market inefficiency.

Let us consider a simple situation: a transaction in which just only one of the parties (who sells) knows about the real quality of the goods. Would you agree to pay more for something you can't be sure about what is being 'promised'? Not existing any other mechanism for ascertaining (measuring) this 'quality' (its key attributes), certainly not. Here we have a situation of informational asymmetry leading to adverse selection. The predominance of sellers and products of suspicious reputation and quality are expected in the market. Theoretical discussion originally published in the 1970 study, that granted to Prof. George A. Akerlof the Nobel Prize in Economics in 2001. See Box *Car for sale*.

From that context, comes the market inefficiency. *"The problems and costs of measurement pervade and significantly affect all economic transactions. Errors of measurement are too costly to eliminate entirely... permitting manipulations;*

The Son of Man, 1946 by Rene Magritte

"At least it hides the face partly well, so you have the apparent face, the apple, hiding the visible but hidden, the face of the person. It's something that happens constantly. Everything we see hides another thing, we always want to see what is hidden by what we see. There is an interest in that which is hidden and which the visible does not show us. This interest can take the form of a quite intense feeling, a sort of conflict, one might say, between the visible that is hidden and the visible that is present".

Source: In Wikipedia, referring to a radio interview (with Jean Neyens cited in Torczyner, Magritte: Ideas and Images, trans. Richard Millen (New York: Harry N. Abrams), p. 172. See at: https://en.wikipedia.org/wiki/The_Son_of_Man#cite_note-3.

Source: Image, personal collection. Drawing made by my son João Francisco motivated by the book published by Springer.

requiring safeguards" (Barzel 1982). In order to solve this economic problem, "(…) *reputation, or brand name, serves to guarantee that the product is, and will remain, uniformly good"* (Barzel 1982). However, this is true specifically when related to commodity or experience goods. Otherwise, market inefficiency remains under threat.

Precisely towards this direction, as discussed in this book, competitive practices in the food industry based on belief goods are spreading. Requiring that, in order to mitigate the informational problems related to its brand strategies, the industry must adopt a new positioning in its value chain, as discussed in Chap. 5. Deriving, further, the premise that harmonizing technology advances with the relationships established by a company and its stakeholders (Silva et al. 2018).

Car for sale
The market for lemons (Akerlof 1970)

- Prob (good car) = 1 − Prob (bad car)
  (Hypothesis: 50%/50%)
- Consumer with reserve prices:

  - for a good car = $ 10,000
  - for a bad car = $ 6000

- Expected value of a car chosen at random:
  $V = 10,000 \times 50\% + 6000 \times 50\% = \$ 8000$
- Considering a risk-indifferent buyer,
  It accepts to pay max $ 8000 for a car of unknown quality
- Salesman who has a bad car, agrees to sell for $ 8000
- BUT, a seller who has a good car, does not accept it and leaves the market.

6.2   The Best of Times

In fact, a clear and crystalline observation, which must be perceptible and visible to all, is that the social responsibility of the contemporary industry is dual: providing safe food for the individual (encompassing nutritional aspects) and for the society (environmental and social issues, in terms of relationship quality along the value chain).

This opens differentiation perspectives based on actions inside and outside the limits of processing technology and the ability above described in terms of the alignment between 'product and process engineering' and 'relationship engineering'.

An effort, however, not spontaneous, or that not should be understood-solely-as an act of organizational detachment. It is, in fact, a strategic business due to the power of the competitive advantage provided. Healthy relationships among the company and its suppliers' reserve authenticity to the product structured in unique attributes and hardly copied by competitors.

But, if on the one hand, differentiators built outside hold the sustainability of competitive advantage, on the other hand, inevitably, they unfold in a more complex competitive game in terms of the organization and governance of the relationships established by the company with its trading partners, particularly with the rural portion.

This is the context in which the food industry is found and what has been tried to illuminate throughout the discussion held in this book. With this exciting scenario for the sector and its professionals, the subliminal message of this book does not agree with the pessimistic view of some of the researchers (according to the discussion that opened this chapter). Different from that, what is practiced here is an optimistic view led by Saguy et al. (2013) and the age of '*enginomics*'.

As discussed in Silva et al. (2018), suggesting the importance of scientific and technological development to the social dynamics in which the industry operates, considering emerging trends such as health and well-being. For these researchers, the attractiveness and relevance of what is yet to come are related to interdisciplinary knowledge, and a new conception in terms of skills expected for the contemporary professional.

We have a huge research agenda that brings many opportunities to rethink the training of human resources in the food industry in general. Ensure solid training in the so-called hard skills (technical and specific domains of professional activity) are critical but can no longer be understood as sufficient (Silva et al. 2018).

As discussed in Silva, Sereno and Sobral (Silva et al. 2018), in the days we are living, we have the responsibility to go further in the technical and economic feasibilities of systems design, addressing an entire spectrum of aspects (Alwi et al. 2014), such as innovation, partnerships, creativity, entrepreneurship, sustainability, economic environment, social responsibility, population growth and aging (Saguy and Taoukis 2017).

Along the trajectory that has been observed since the conception of the contemporary industry, the sector has been marked by cycles of knowledge development increasingly narrow. But this is not everything. Bringing more value to the discussion, it cannot be underestimated that the opportunity of the food professional also lies in being familiar with interdisciplinary knowledge, such as medicine, molecular gastronomy, nanotechnology, novel materials, among other spheres of the scientific domain (Silva et al. 2018). Including fundamentals of economy and business, and for instance economic tools of market analysis and relationship governance.

> **Looking out of the box**
> All this repositioning that was seen impacts on the contemporary training of the professional in the food area in different ways. New domains and approaches are expected, in the same way as an update of the classic themes. Take for example quality standards. The classic ISOs are no longer sufficient, although duly necessary. But an urgent renewal of standards is expected, as Prof. Humphrey invites us to think in the report 'Global value chains in the Agri-food sector' (Humphrey and Memedovic 2006). Aligned to this understanding, the food ethics differential projects several and important new standards of quality, both public and private, attesting to social and environmental attributes. Such as 3 A (Nespresso), Rainforest, Company B, etc.

Unlike the usual understanding of the area, relationships are not data (*ad hoc* solution). Relationships must be built, and they are built in the corporate world where the food industry stands and must position itself as competitive.

A challenge that expands due to supplier and time constraints. Therefore, contribute to the formation of better prepared human resources, addressing the challenges that the industry poses, requires instigating them to develop soft skills.

And that's not all. It is necessary to be familiar with other abilities, such as communication, project management and group work. The contemporary food professional should also be encouraged to deal with conflicting situations, developing the skills to leave the comfort zone, obtaining results in an uncontrolled atmosphere, as well as resilience capacity in order to confront the pressure of different natures, such as deadlines.

Summing up, the thought-provoking history of the food processing industry and its professionals is far from the end. And it was sought to discuss here that the best chapters reserve the best of times.

Assuming the statement of Saguy et al. (2013): *"the best of times, the worst of times"*.

The sector has to reinvent itself all over again. Going out of the box and positioning itself to take advantage of the opportunities that exist outside; while we, educators, also face the urgent demand for reconstructing professional training.

References

Aguilera, J. M. (2006). Perspective Seligman lecture 2005. Food product engineering: Building the right structures. *Journal of the Science of Food and Agriculture, 86*, 1147–1155.

Akerlof, G. A. (1970). The market for "lemons": Quality uncertainty and the market mechanism. *The Quarterly Journal of Economics, 84*(3), 488–500. https://doi.org/10.2307/1879431.

Alwi, S. R. W., Manan, Z. A., Klemes, J., & Huisingh, D. (2014). Sustainability engineering for the future. *Journal of Cleaner Production, 71*, 1–10.

Barzel, Y. (1982). Measurement cost and the organization of markets. *Journal of Law and Economics, 25*, 27–48.

Humphrey, J., Memedovic, O. (2006). *Global value chains in the agrifood sector*. Retrieved from http://tinyurl.com/y8qh4obd.

Kasemodel, M. G. C., Makishi, F., Souza, R. C., & Silva, V. L. (2016). Following the trail of crumbs: A bibliometric study on consumer behavior in the food science and technology field. *International Journal of Food Studies, 5*, 73–83.

Lopez, J. (2014) Changing role of business. Creating Shared Value Forum 2014. Retrieved from https://www.youtube.com/watch?v=SOz5Qw1kc84.

Mintel. (2016). *Global food and drink trends 2016*. Retrieved from https://www.mintel.com/presscentre/food-and-drink/mintel-identifies-global-foodand-drink-trends-for-2016

Saguy, I. S., Singh, R. P., Johnson, T., Fryer, P. J., & Sastry, S. K. (2013). Challenges facing food engineering. *Journal of Food Engineering, 119*, 332–342. https://doi.org/10.1016/j.jfoodeng.2013.05.031.

Saguy, S., & Taoukis, P. S. (2017). From open innovation to *enginomics:* Paradigm shifts. *Trends in Food Science & Technology, 60*, 64–70. https://doi.org/10.1016/j.tifs.2016.08.008.

Silva, V. L., Makishi, F., Magossi, M., Moraes, I. C. F., Trindade, C. S. F., & SOBRAL, P. J. A. (2018). Are we doing our homework? An analysis of food engineering education in Brazil. *International Journal of Food Studies, 7*, 1–16.

Silva, V. L., Sereno, A. M., & Sobral, P. J. A. (2018). Food Industry and processing technology: On time to harmonize technology and social drivers. *Food Engineering Reviews, 10*, 1–13. https://doi.org/10.1007/s12393-017-9164-8.

Singh, R.P. (2012). Romancing with food engineering: A life-long second partner. 2012. Annual Meeting of the Institute of Food Technologists, Las Vegas. Retrieved from https://www.youtube.com/watch?v=23eqfw2aaI8.

Weaver, C. M., Dwyer, J., Fulgoni, V. L., III, King, J. C., Leveille, G. A., MacDonald, R. S., Ordovas, J., & Schnakenberg, D. (2014). Processed foods: Contributions to nutrition. *American Journal of Clinical Nutrition, 99*, 1525.

Index

© Springer Nature Switzerland AG 2020
V.-L. Silva, *Social Drivers In Food Technology*,
https://doi.org/10.1007/978-3-030-50374-1

Printed in the United States
by Baker & Taylor Publisher Services